The author, Marin County, California. Photo taken by Elke May Glendenning.

Norman K. Glendenning took his bachelor and masters degrees at McMaster University in Canada, and PhD in theoretical physics at Indiana University in 1959. Glenn T. Seaborg invited him to continue his research at the Lawrence Berkeley Laboratory in 1958. There he has spent his entire professional career aside from frequent short-term visits of a month to a year as visiting professor in Paris and Frankfurt. He won the Alexander von Humboldt prize in 1994.

He has published 273 scientific research papers in scholarly journals and in the proceedings of international conferences as invited lecturer — in North and South America, Europe, Asia, and India. As well, he has published a technical book in each of the two major fields in which he has done research over the years, *Direct Nuclear Reactions* (Academic Press, 1983), and *Compact Stars* (Springer-Verlag, 1st ed., 1997, 2nd ed., 2000). A technical book on *General Relativity* is being published by Springer, New York.

More recently, he studies — as a student — the work of others, and writes what he has learned for lay readers, students, and scientists in other fields — *After the Beginning: A Cosmic Journey through Space and Time.* This book was selected by Scientific American Book Club as an alternate selection in July 2005. The present book, *Our Place in the Universe* is his fifth book — also for the lay reader and for use in a general college course on the solar system.

Norman K. Glendenning

Lawrence Berkeley National Laboratory

World Scientific

Imperial College Press

Published by

Imperial College Press
57 Shelton Street
Covent Garden
London WC2H 9HE

and

World Scientific Publishing Co. Pte. Ltd.
5 Toh Tuck Link, Singapore 596224
USA office: 27 Warren Street, Suite 401-402, Hackensack, NJ 07601
UK office: 57 Shelton Street, Covent Garden, London WC2H 9HE

Library of Congress Cataloging-in-Publication Data
Glendenning, Norman K.
Our place in the universe / Norman K. Glendenning.
p. cm.
Includes bibliographical references and index.
ISBN-13 978-981-270-068-1 -- ISBN-10 981-270-068-4
ISBN-13 978-981-270-069-8 (pbk) -- ISBN-10 981-270-069-2 (pbk)
1. Cosmology. 2. Galaxies--Formation. 3. Galaxies--Evolution.
4. Astrophysics. 5. Solar system--Origin. I. Title.

QB981 .G585 2007
523.1--dc22

British Library Cataloguing-in-Publication Data
A catalogue record for this book is available from the British Library.

Cover picture:
The Horsehead is a reflection nebula; it is located in the much larger Orion Nebula, an immense mocleular cloud of primordial hydorgen and helium, together dust cast off by the surfaces of stars. The clouds that form the Horsehead are illuminated from behind by a stellar nursery of young bright stars.
Credit: Photo taken at Mauna Kea, Hawaii with the Canada-France-Hawaii Telescope.

First published 2007
Reprinted 2008

Printed by FuIsland Offset Printing (S) Pte Ltd, Singapore

I dedicate this book with love to my family:

my son Nathan
his mother Laura
and
Elke and Alan.

Greek Creation Myth: In the beginning there was an empty darkness. The only thing in this void was Nyx, a bird with black wings. With the wind, she laid a golden egg, and for ages she sat upon this egg. Finally, life began to stir in the egg and out of it arose, Eros, the god of love. One half of the shell rose into the air and became the sky and the other became the earth. Eros named the sky Uranus and the Earth he named Gaia. Then Eros made them fall in love. Credit: Lindsey Murtagh (her website was completed for a high school Latin course).

Preface

Our small place in space — this planet Earth — together with Sun and other planets of our solar system and their many moons is located about two-thirds from the center of the disk-shaped galaxy that we know as the Milky Way. When we look toward its center on a clear night what we see is the faint luminosity of billions of stars that make up this galaxy.

There are other galaxies too, far away, that look to our naked eyes just like any other star. But they too contain billions of stars. From our vantage point on Earth, orbiting a rather insignificant star that we call Sun, located in the Milky Way Galaxy in one of its many solar systems — we gaze in wonder.

At last, I have come to that stage in life and career that I can put aside the urge to understand some little details, however fundamental they may be in the edifice of a particular science — in my case nuclear and astrophysics — and look to the grander scale of our place in the universe with its Sun and planets and their many moons stretching far away. I tell, too, the story of life, its needs, and vicissitudes on Earth and in its seas, as well as other possible habitable zones on other planets and on the larger moons of some of them. It is a story that often as children we yearn to comprehend, and concerning which, as adults, we may achieve a degree of understanding.

I write with the urge of writing, because it makes me happy. I step eagerly to my desk in the morning to research what the experts in many special fields of science have learned. I try to understand what they so painstakingly have researched over the decades, making sense of the pieces and then synthesizing what I have learned with some poesy I hope, to please myself and the reader like me — not an expert. In other words I write for the general lay reader, college students taking a general science course, as well as scientists in other fields. And in the evening, tired, I leave, only to take up where I left off the next day. Some days though, when the Sun shines warmly I leave early; we were made to love the Sun and its life-giving rays of energy. Sunday is my day of rest from study and writing.

Norman K. Glendenning
Lawrence Berkeley National Laboratory
April 9, 2004

A faint nebula in the northern hemisphere is associated with the bright star Rigel in the constellation of Orion. The Witch Head Nebula glows primarily by reflected light. Credit and permission: John C. Mirtle.

Acknowledgments

I am deeply indebted to my dear friend, Joerg Huefner, Professor Emeritus at the University of Heidelberg. He has read most of the chapters throughout their writing and many rewritings. I have greatly enjoyed the many emails we have exchanged, both of a technical and a personal nature. To the extent that the book is coherent and self-contained now, I owe him great thanks, which I record here.

I am grateful to Dr. Viola Ruck, Professor at North Lake College, Irving, Texas, my one-time valued research associate here in Berkeley, for her encouraging words for my previous book, *After the Beginning*, and her generosity in reading the present one. She has proofread the manuscript with great care and called my attention to innumerable details which needed attention. I owe her special thanks for this careful work: it was a great favor and the book owes much to her.

To Laura Louis for her never-ending generosity, support and encouragement, I say thank you from the bottom of my heart. And to our dear and fine son Nathan I dedicate this book with love.

Any acknowledgment would be incomplete if it did not pay tribute to the National Aeronautics and Space Administration (NASA), its personnel, and the many successful missions that have so enriched our knowledge of the universe at both the scientific and cultural levels.

Contents

List of Figures

Chapter 1

A Day without Yesterday

Man is equally incapable of seeing the nothingness from which he emerges and the infinity in which he is engulfed.

Blaise Pascal, *Pensées* (1670)

Not only is the universe stranger than we imagine, it is stranger than we can imagine.

Sir Arthur Eddington (1882–1944)

1.1. Instant of Creation

In an instant of creation about 14 billion years ago the universe burst forth, *creating* space where there was no space, and time when there was no time. It was so hot and so dense that not even nucleons, the building blocks of atomic nuclei, had as yet formed. Until about 1/100,000th of a second all that existed was an intense fire and primitive particles called quarks and electrons and all their heavier kin together with their antiparticles; in a sense, these were the elementary particles conceived of by two early Greek philosophers of the fifth century BC, Leucippus and his disciple, Democritus. They evidently believed that an end must come to the reduction of matter into smaller parts, which they called *atomos*. Indeed, elementary particles that we call *quarks* have been discovered at last in our own time and all attempts to reduce them further have failed.

After that first brief moment, the quarks coalesced into neutrons, protons and other versions of these nucleons, never to be free again. In that inferno the lightest elements — deuterium and helium — were forged from the neutrons and protons during the next few minutes. These two *primordial elements* make up fully one-quarter of the mass in the universe today. Almost all the remaining mass is in that simplest element — hydrogen — consisting of one proton orbited by one electron. The hydrogen nuclei that were formed in the first fraction of a second are the same that pervade the universe now.

In the very earliest moments, there must have been a slight lumpiness in an otherwise uniformly expanding universe. As the expansion progressed, gravity acted upon those lumps, attracting surrounding matter — the primordial hydrogen and helium — into great tenuous clouds of chaotically swirling gases and the radiation that was trapped inside them. As the radiation slowly leaked from their surfaces and the clouds cooled, they began to collapse and fragment to form proto-galaxies, some slowly turning in one direction, others in another, so that in sum there was no rotation.

As the proto-galaxies collapsed further, conservation of their angular momentum caused them to whirl ever faster, just as an ice skater — who goes into a whirl with arms outstretched and draws them in — whirls faster. As each cloud fragment collapsed, it was flattened like a pancake with a central bulge by the increasingly rapid rotation, like the Andromeda Galaxy shown in Figure 1.1. At last a balance was reached between the force of gravity that would cause total collapse, and the centrifugal force of rotation, that would spin matter off into space. Because of instabilities within the thin disks, the clouds fragmented further to form stars, some larger than our Sun, and many more of them smaller.

When galaxies were first observed in the first century as faint nebulous objects in the sky it was not known that they were enormous collections of stars (Figure 1.2). Al Sufi was an astronomer in the court of the Emir in Persia. He observed the great galaxy, Andromeda, as a faint nebulous patch, which he called "little cloud" in his famous *Book of Fixed Stars* in 964 A.D. And he described other *nebulae*, some of which are not galaxies at all, but rather dense clouds of gas and dust in our own galaxy, the Milky Way. The Witch Head Nebula, shown in the preface, is one of many beautiful nebula, some of which have an appearance that reminds us of an animal, or of a mythological figure.

Not until the late 1700s did William Herschel speculate that some of the hazy patches that he could see among the stars with his telescope were actually "island universes" like our own galaxy, lying far outside it and containing vast numbers of stars. For his accomplishments, William Herschel was knighted by George III of England. His sister, Caroline, too (Figure 1.3) made many important discoveries and was the first woman to be recognized as an astronomer with a salary from the King. By the end of her long life she had reaped numerous honors including the gold medal of the Royal Astronomical Society (England).

Fig. 1.1 The Andromeda Galaxy (M31), was first recorded by the Persian astronomer Al Sufi (903–986 A.D.) living in the court of Emir Adud ad-Daula. He described and depicted it in his *Book of Fixed Stars* (964 A.D.) and called it the "little cloud." The galaxy is composed of about 400 billion stars. It lies relatively close to our own galaxy, the Milky Way, and is rather similar, having a central bulge and flat disk in the form of spiral arms. Relatively close in this context means 2,900,000 light-years (or equivalently a thousand billion kilometers). The two galaxies are attracting each other and will collide and pass through each other, causing distortion of each. Eventually, they will merge. Two smaller elliptical galaxies M32 and M110 (the two bright spots outside the main galaxy) are in orbit about Andromeda. The myriad foreground stars are in our own galaxy. Credit: George Healy obtained this color image of spiral galaxy M31, the Great Galaxy in Andromeda, together with its smaller elliptical satellite galaxies.

But it was not until the 1900s that astronomers using more powerful telescopes were actually able to discover that the Milky Way contains about four hundred billion stars. It is a giant among galaxies having a mass of about 600 billion to a trillion times the mass of our Sun; it is exceeded in our neighborhood only by the Andromeda Galaxy, which is generally regarded as the most distant object visible to the human eye. At a distance of 2.2 million light-years, it appears as a fuzzy patch of light in the night sky.

Fig. 1.2 Many galaxies, such as the one in Figure 1.1 are seen in this deep view of distant galaxies. Some are similar to our own, the spiral galaxies, seen at various angles. Others are spheroids or ellipsoids, called elliptical galaxies. Credit: NASA, ESA, and the Hubble Heritage Team (AURA/STScI).

But only the bright core of the galaxy can be seen by the eye. The full extent of the galaxy covers over 3 degrees of sky on its longest side. The Milky Way is falling towards our nearest large neighbor, the Andromeda Galaxy, M31, attracted by its great mass. In about 10 billion years the two systems will collide and merge. The end product of this merger is likely to be an elliptical or spherical galaxy with a very high concentration of stars and gas at its center, and very diffuse beyond, gradually fading into nothingness.

Fig. 1.3 Caroline Herschel (1750–1848), winner of many awards for her work in astronomy including the Gold Medal of the Royal Astronomical Society (England), and the Gold Medal of Science by the King of Prussia, first woman to be recognized and salaried as a scientist by King George III.

Ten percent of galaxies are estimated — based on modern observations — to possess planetary systems containing large Jupiter-like planets, and no doubt many others too small to detect, perhaps some like Earth. Besides our own Sun and its planets, there are 133 other known planetary systems in our galaxy, containing at least 156 planets in orbit around main sequence stars. It is very probable that there are many smaller planets in these systems, but it is easiest to detect massive planets because their effect on the host star is more apparent. Consequently, this number increases from year to year as more sensitive detection instruments are put into play.

At the center of our galaxy, as appears to be the case with others, lies an enormous black hole that is ingesting stars. It is hidden from our direct view by dust and surrounding stars spinning about it ever faster as they fall toward their fate. Our own Sun is far removed from this cannibalism. It lies in the disk of our Milky Way about two-thirds from its center. Although the Sun is moving at about 250 kilometers a second, the circumference of its orbit is so large that it has circled the center only 18 times since its birth 4.5 billion years ago.

The first generation of stars appeared several hundred million years *after the beginning.* They were supermassive — several hundred to a thousand times more massive than our Sun — and short-lived. Their intense gravity, owing to their large mass, soon crushed them, destroying atomic nuclei and sequestering a fraction of their nucleons inside a new star — a neutron star or even a black hole. Such events release enormous kinetic energy in a cataclysmic explosion called a *supernova.*

The most famous supernova was reported in 1054 to the emperor of the Sung dynasty by the imperial astronomer:

> *"I bow low. I have observed the apparition of a guest star. Its color was an iridescent yellow . . . ."*

Fearing the mighty emperor, he added a favorable interpretation: "The land will know great prosperity." This supernova was visible in daylight for 23 days in China. Anasazi Indian artists (who lived in present-day Arizona and New Mexico) created a pictograph that is thought to commemorate this amazing event; it accurately locates the position of the supernova in the night sky relative to the crescent moon which would have been its phase on the date of the explosion's appearance (Figure 1.4). Japanese historical records also noted the sudden appearance of a guest star.

This explosion, supernova 1054, created the Crab Nebula (Figure 1.5) that will be visible for thousands of years. It is a much-studied object, for it houses within it a *pulsar*, a neutron star with a mass nearly as much as our Sun; it is rotating 33 times a *second.* A very strong magnetic field is fixed in the star, which, because of the rotation, is pumping out energy like a dynamo with a luminosity of 100,000 Suns. This energy illuminates the Crab Nebula, and as well, has accelerated wisps of gas at its periphery to the very high velocity of about 1,500 kilometers per second.

The last star *observed* to explode in our own Galaxy produced the beautiful nebula called Cassiopeia in about 1667 (Figure 1.6). It is named for a mythological queen, the wife of Cepheus and the mother of Andromeda. Cassiopeia was a vain woman who thought herself more beautiful than the daughters of Nereus, god of the sea. To teach her humility, Cassiopeia was banned to the sky as a constellation of stars, hanging half of the time head downward (Figure 1.7).

The very light elements were made in the intense heat and high density that existed only in the first few minute in the life of the universe. The first traces of the all important elements for life — carbon, oxygen, and iron — were forged later in the first generation of stars. They were supermassive

Fig. 1.4 On July 4, 1054 A.D., inhabitants of Earth noticed a bright object in the sky, easily visible even in daylight — the supernova of 1054. This was the supernova which created the Crab Nebula. At that time the moon was a small crescent. Archaeologists believe the Anasazi Indians recorded the supernova event with this rock art panel in Chaco Canyon, New Mexico. Photo credit: Ron Lussier.

and appeared after several hundred million years. And the molecular dust that they cast off from their ferociously burning surfaces formed the seeds about which whole galaxies containing billions of stars eventually condensed.

The other elements were not made in significant abundance until later generations of stars — up to a few times our Sun's mass — first made their appearance. Such stars forged, and are forging, elements up to iron in their interiors but release them only upon their deaths in supernova explosions. Heavier elements like nickel and beyond are synthesized outside these stars when an intense wind of neutrinos created at the star's final collapse under gravity irradiates the hot gases that are expelled from the dying star.

In far-off galaxies, even today, supernova explosions occur, ejecting the store of elements that the exploding stars had synthesized during their lifetimes. Following countless of these starry deaths in our own galaxy, this

Fig. 1.5 The Crab Nebula, created by the supernova explosion of a star in 1054 and recorded by the court astronomer in China. The supernova remnant is expanding at 1,800 kilometers per second. It is about 10 light-years across. The source of the illumination is a rapidly rotating *neutron star* which has the luminosity of 100,000 Suns. Credit: European Southern Observatory/Very Large Telescope.

atomic debris has wandered for eons — mixing with that of other exploding stars — finally to coalesce to form our solar system — sun, planets, moons — and we, ourselves.

1.2. Size and Age of Earth, Sun and Milky Way

Our own experience of seeing objects in everyday life is as if light were an instantaneous messenger. For all practical purposes this is so. Light travels at the speed of 300,000 kilometers per second. But the distance between galaxies is so vast that a different measure than a mile or kilometer is needed. The unit that is useful to us here is the distance that light travels in a *year*. This unit is called the *light-year*; it is approximately 10,000 *billion*

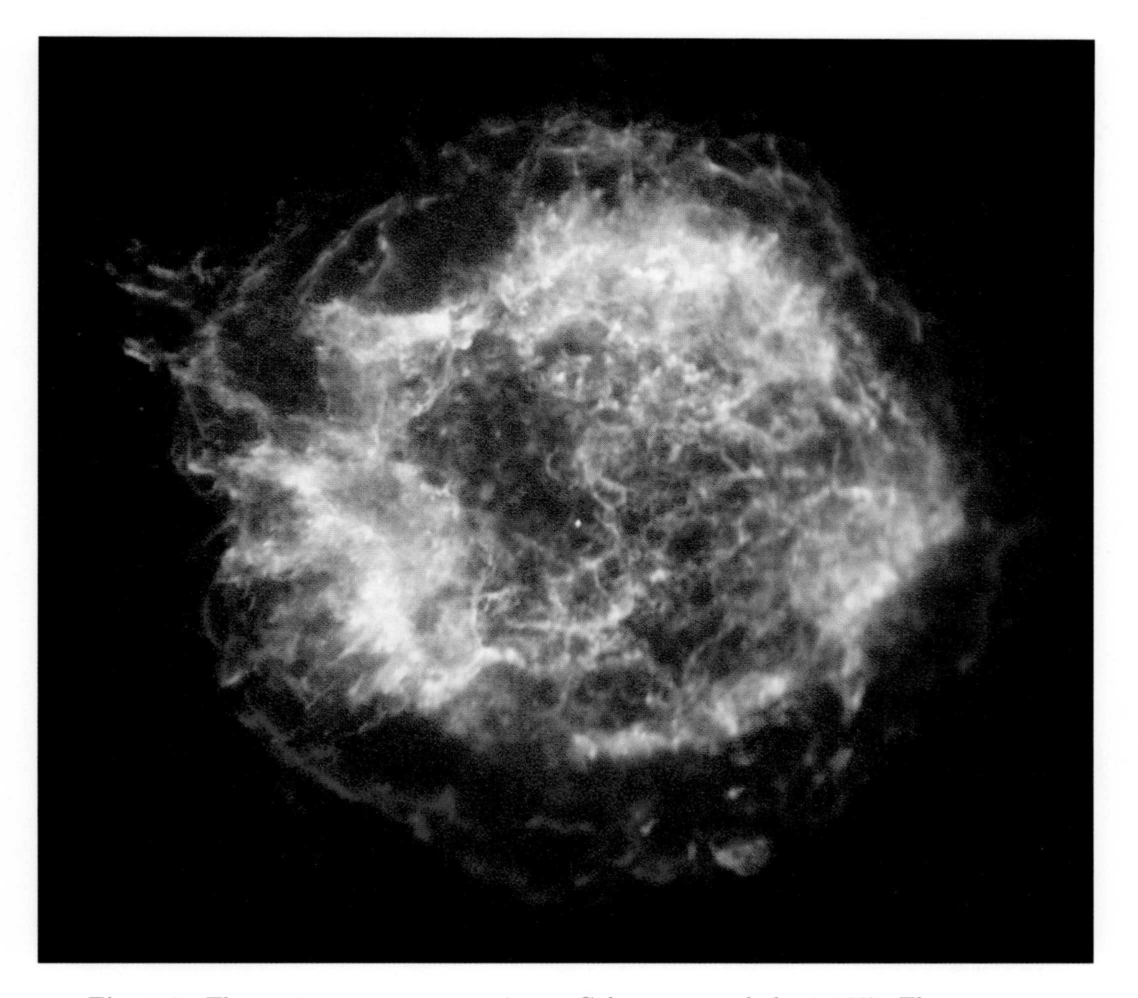

Fig. 1.6 The most recent supernova in our Galaxy occurred about 1667. The remnant, called Cassiopeia — now about 15 light-years in diameter — is seen here in x-rays in this Chandra X-Ray Observatory photo. Credit: NASA/CXC/MIT/UMassAmherst/M.D. Stage *et al.*

kilometers. A far-off galaxy is seen *now* as it was in the past by the light-travel-time from it to us. For example, the Large Magellanic Cloud — the nearest galaxy to us — is 169,000 light-years away. The light we see from it *now* has taken 169,000 years to reach us. Useful astronomical measures are given in Box 1 along with the age of the Earth, Sun, and other interesting data. (See page 30.)

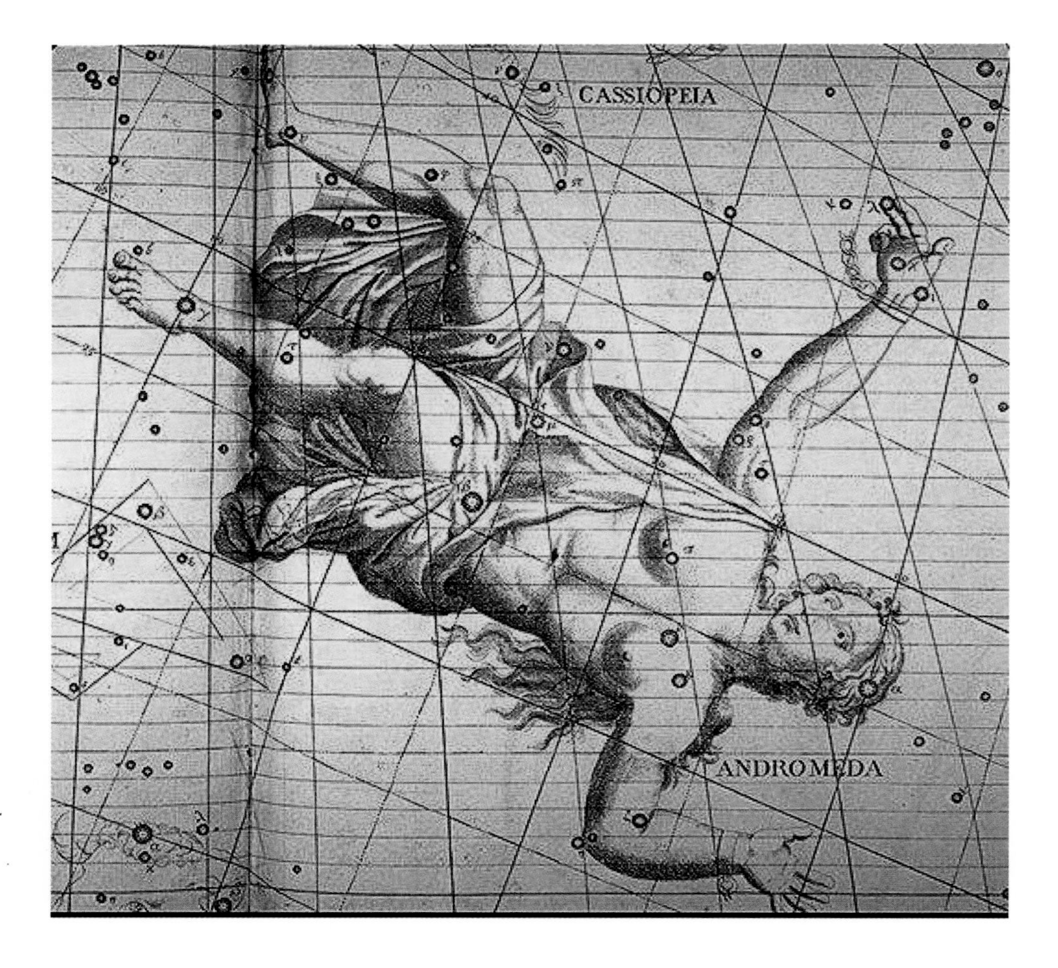

Fig. 1.7 Cassiopeia was banned to the sky as a constellation of stars, hanging half of the time head downward. This illustration of the constellation is from a book by John Flamsteed (1645–1719).

The Milky Way Galaxy in which our Sun resides, among about 12,000 billion others, is very old, about 10 billion years. Compared to our own perspective on Earth it is hard to discern with our senses, let alone comprehend with our minds the vastness of space and the sizes, ages, and masses of the various heavenly objects.

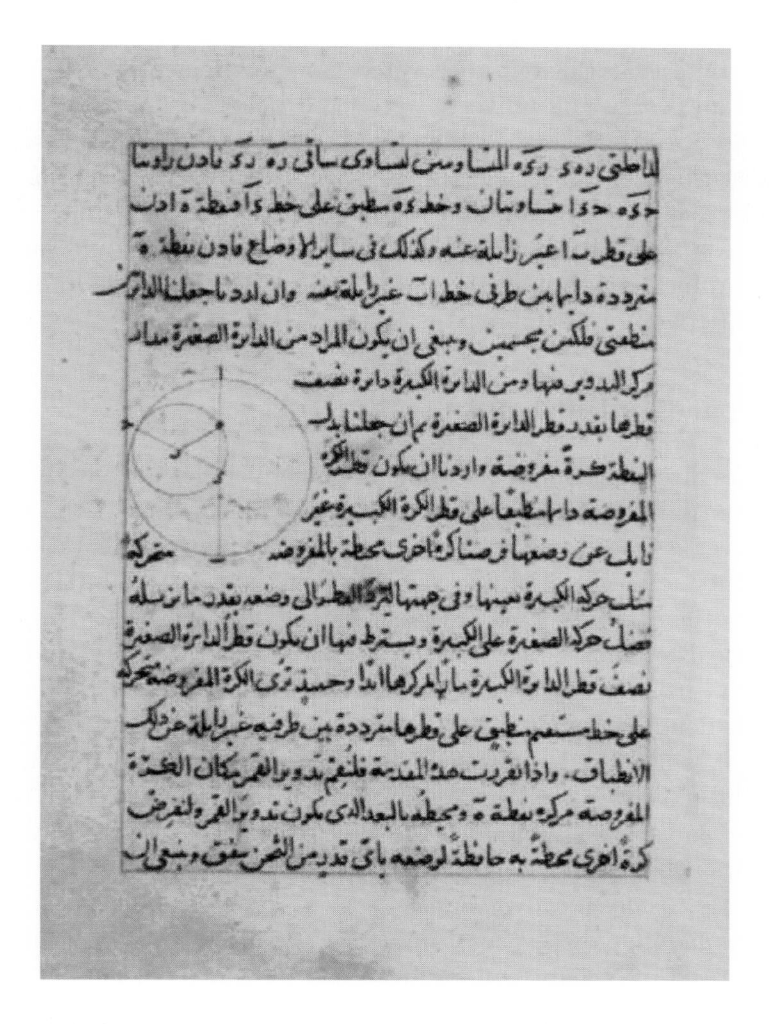

Fig. 1.8 Nasir ad-Din al-Tusi (1201–1274) was among the first of several Arabic astronomers at the observatory of Maragha in Persia who modified models based on mechanical principles by Ptolemy (87–150). This early Arabic manuscript contains the *al-Tadhkira fi ëilm al-hayía* (Memoir on Astronomy). This scholar wrote on many subjects besides astronomy, including poetry, mathematics, medicine, history, law and philosophy.

1.3. Einstein's Static Universe

> *Much later, when I was discussing cosmological problems with Einstein, he remarked that the introduction of the cosmological term was the biggest blunder of his life.*
>
> George Gamow, *My World Line*

At the time that Einstein discovered General Relativity (1915), permanence and an everlasting universe were fixed beliefs in Western philosophy. Hubble had not yet discovered that the universe is expanding. Einstein was greatly disturbed that his theory, as it stood, described a changing universe. He therefore added what he named the cosmological constant — denoted by the Greek symbol Λ (lambda) — so that his theory could be made to describe an unchanging universe that would neither collapse under its own gravity, nor coast forever in expansion (Box 3).

1.4. Expanding Universe

A few years after Einstein's discovery of General Relativity, Edwin Hubble, an astronomer in California, shattered the age-old belief in the constancy of the universe. He discovered through his observations that the galaxies around us are actually speeding away, and what is more, the further they are, the greater their speed. This discovery came to be known as Hubble's law. Mathematically it has a very simple expression; the velocity v of a galaxy moving away from us is proportional to its distance d and the proportionality constant is denoted by H_0, after Hubble; the relationship is $v = H_0 d$. The subscript, 0, is used to denote the *present* value of the proportionality factor, because the universe is not expanding at a constant speed. For billions of years the expansion was slowed by the force of gravity, but for the last several billion years it has been accelerating because of whatever mystery is covered by the words *cosmological constant.*

Hubble made his discovery of universal expansion by measuring what is called the redshift of light coming from nearby galaxies. The redshift, also commonly called the Doppler shift, is well known to anyone who has heard the high pitch of the whistle of an approaching train sink to a lower pitch as it passes. All wave-like motion, including light, will experience this sort of shift if it originates from a moving source. Light from a star or galaxy that is receding will appear redder: it is said to be redshifted. By measuring the redshift of other galaxies beyond our own, Hubble made his amazing discovery (Box 5).

According to our present understanding concerning the expansion, what was originally called the Doppler shift is more accurately named the *cosmological redshift.* The Doppler *redshift* refers to the elongation of light wavelengths of radiation received by an observer from an object moving away from him, and both within an *unchanging* space. The cosmological redshift of radiation from distant galaxies is caused by the elongation of light wavelengths in the *expanding* space of the universe in which the galaxies are *co-movers* in that space.

As the universe expanded, radiation, such as the light and heat we receive from the Sun, cooled in inverse proportion to the expanding size of the universe (Box 8). This we know from simple and basic laws of nature. How fast it expanded is also governed by the laws of nature, but to put numbers to it, we need to know a few — indeed a very few — important constants called the *cosmological* constants. (The three epochs of expansion can be seen in Box 7.) Their values have been discovered only during the last decade.

In the early universe, matter was in such close contact with radiation that it too cooled with the expansion in the same way as radiation until about 300,000 years. After that the primordial radiation streamed through the universe as if nothing else were there because the number of photons of light so vastly outnumbered charged particles (Box 9). We can deduce this from the discovery of Penzias and Wilson. They were trying to improve satellite communications for Bell Laboratories in New Jersey when, in 1965, they discovered something very unexpected — a faint radio signal that was coming from *all* directions of the sky.

This omni-directional radiation, known as the *Cosmic Background Radiation* (CMBR), seemed quite puzzling, for it could not be arriving from other galaxies; they are localized dots in the sky when viewed from earth. Moreover, the radiation was very cold, about 270 degrees centigrade *below* the freezing point of water. George Gamow had foreseen the likely existence of this whisper from the past. But, unaware of his foresight, Robert Dicke and Jim Peebles, two Princeton scientists, had come to the same conclusion and were actually designing receivers and antennae to detect it when word reached them of its discovery. "We've been scooped" Dicke murmured to his colleagues when a telephone call reached him with the news. This discovery won Penzias and Wilson the 1978 Nobel Prize: It was the first very convincing evidence of the hot and dense beginning, sometimes known as the *Big Bang*[a] that had been proposed, in different languages, by the Belgian theological student, Georges Lemaître in 1927, and independently by the Russian meteorologist and bomber pilot, Alexander Friedmann. Lemaître referred to the beginning as "*a day without yesterday.*"

[a]This rather inelegant phrase for the sublime moment of the beginning of the universe was coined by Fred Hoyle, a well-known astronomer and cosmologist, on a BBC radio broadcast. That phrase is not used again in this work.

Both Friedmann and Lemaître had read Einstein's relativity papers and what was more, somehow understood without there being any evidence of it at the time that from a very dense and hot beginning, the universe burst forth to expand possibly forever. (Box 2 and Box 7.)

This is the most profound and natural implication of Einstein's General Relativity paper (that and the existence of black holes). Einstein himself had not realized this, thinking, as his contemporaries and Newton before him, that the universe was forever unchanging.

The young Lemaître, later a priest, was struggling at the time (1932) to reconcile the biblical account of creation with Einstein's theory of gravity (General Relativity). To further his task, he took up the study of astrophysics and cosmology at Cambridge University and the Massachusetts Institute of Technology. He published his early ideas on the birth of the universe in an obscure Belgian journal where it passed unnoticed by the principals in the field. Lemaître applied Einstein's theory of gravity to cosmic expansion and conceived the notion of the "primeval egg", which is the term he used to describe the universe at its beginning. His prescient notion of an eventual cosmic acceleration, which he included among the possible universes, has actually been confirmed only in the past few years and ranks among the great cosmological discoveries of all time. Friedmann too, working independently in Russia, came to the same conclusion.

It is all the more remarkable that Lemaître's publication of his audacious theory of the beginning of the world actually preceded Hubble's 1929 momentous discovery of universal expansion. After he had learned of Hubble's discovery, he quoted that work in his own subsequent publications to support his ideas. Meanwhile, Hubble, in California, was unaware of the young priest's theory, which had foreshadowed his own great discoveries.

Friedmann submitted his paper with the surprising conclusion that the universe may be expanding to the German physics journal *Zeitschrift für Physik* whose editor sought Einstein's advice as a referee. He wrote back saying; "The results concerning the non-stationary world, contained in [Friedmann's] work, appear to me suspicious. In reality it turns out that the solution given in it does not satisfy the field equations [of General Relativity]."

However, Friedmann was confident of the results that he had obtained from Einstein's theory. He wrote to Einstein beginning: "Considering that the possible existence of a non-stationary world has a certain interest, I will allow myself to present to you here the calculations I have made ... for verification and critical assessment...." Meanwhile, Einstein had already

Fig. 1.9 Albert Einstein and J. Robert Oppenheimer discuss their work at the Princeton Institute for Advanced Study. By this time Einstein has long been a world figure for his paper on General Relativity, published in 1915, his name synonymous with genius. It is the greatest single contribution to our understanding of gravity, and is the means by which we can trace the history of the universe and its possible futures. Credit: Emilio Segrè Visual Archives.

left for Kyoto and did not return to Europe for several months. Then, by chance a friend of Friedmann's met Einstein at Ehrenfest's house in Leiden and described his colleague's work; Einstein saw his error and immediately wrote to the journal's editor "... my criticism [of Friedmann's paper]... was based on an error in my calculations. I consider that Mr. Friedmann's results are correct and shed new light."

Ultimately, Albert Einstein was so impressed by the work of Friedmann and Lemaître that at a meeting of Scientists in Pasadena in 1933, he rose after Lemaître's speech to say "This is the most beautiful and satisfactory explanation of creation to which I have ever listened."

1.5. Universe without Center

Edwin Hubble did not know of the theoretical work of Friedmann and Lemaître, when he discovered that the universe is expanding — that distant galaxies are rushing away — and the further they are, the faster. Moreover, he discovered that the number of galaxies increased uniformly within every angular patch of sky the deeper he looked into space, and that this was true no matter the direction. These *large-scale* properties of the universe are referred to as *homogeneity* and *isotropy*, or simply as *uniformity*. The observed *large scale* uniformity has momentous implications.

On the small scale — astronomically speaking — the nearby universe is anything but uniform. We see the Milky Way as a faint band of light across the sky. Nearby stars, those in our own galaxy, are concentrated, either in the disk or else in the central bulge; our Sun is in the disk about two-thirds from the center. In appearance, the Milky Way is similar to the Andromeda Galaxy (Figure 1.1). Even further out in space astronomers see stars concentrated in other galaxies — not randomly scattered about. And further still, galaxies are gathered into *galaxy clusters*. Gravity has had time since the beginning to rearrange the once diffuse matter from which these objects are made.

But on a much grander scale, the astronomer sees galaxy clusters at a distance of a billion and more light-years — which still corresponds to billions of years after the beginning — scattered about reasonably evenly so that what lies at a very great distance in our universe seems much the same in every direction. From these observations we learn that the universe will appear the same to an observer, *on the grand scale*, in any other galaxy as it appears to an observer in ours. This observation of *homogeneity* and *isotropy* of the universe justify what is referred to as the *Cosmological Principle*.

This principle is very important; it allows us to extend our observation of that part of the universe that we can see with our telescopes to the universe as a whole. It has *another* very interesting consequence: Any observer, located anywhere in the universe would see galaxies rushing away from him just as Hubble observed them to be rushing away from us. Therefore he might conclude that *he* was at the center. Clearly the only resolution to this paradox is that *the universe has no center*: As viewed from *any* galaxy, all distant galaxies are rushing away with a speed in proportion to their distance.[b]

[b]By a simple example found in the Section "Questions (number 5)" one can easily perceive why Hubble's observations must be true in an expanding universe.

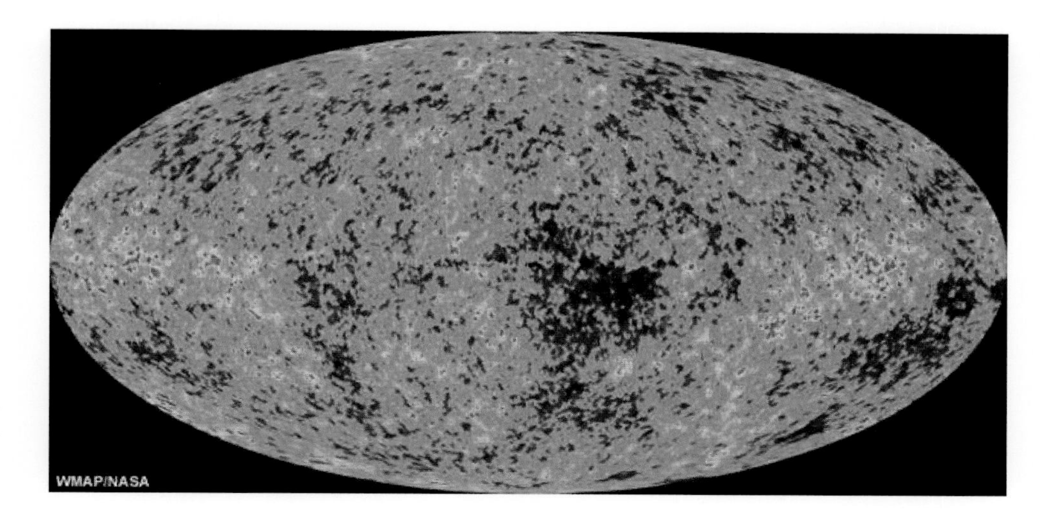

Fig. 1.12 The snapshot shows the state of the universe about 380,000 years after the beginning. No stars were yet present. The radiation present at that time is not perfectly smooth, but lumpy. This lumpiness caused a corresponding lumpiness in the distribution of matter. These clumps formed the seeds of galaxies. The universe is composed of 4% ordinary matter, 23% of an unknown type of dark matter, and 73% of a mysterious dark energy. This is a confirmation of the so-called concordance lambda-CDM model. The study of this cosmic microwave background radiation (CMBR) was made using NASA's space-based Microwave Anisotropy Probe (MAP) observatory. Credit: NASA/WMAP Science Team.

few years by measurements of the distance and age of far off (type Ia) supernovae; these measurements indicate acceleration and a flat ($k = 0$) universe. (See S. Perlmutter, *Physics Today*, April 2003.)

14. *How could astronomers learn that the expansion of the universe is accelerating?*

The expansion of the universe is affected by how much mass density is in it (like stars and galaxies). The more mass density, the more gravity will slow the expansion. In the simplest cosmological model (no mass, no cosmological constant) the expansion would proceed at a steady rate. But there is mass — the galaxies have mass. So the rate of expansion at the location of distant galaxies would be, by this expectation, smaller. It is not: it is *larger*! This implies that the expansion at some time in the past began to accelerate. This is the evidence for a positive cosmological constant Λ.

15. *What actually is measured in the determination of cosmic acceleration?*

It is based on a measure of redshift z. The wavelength of spectral lines coming from the supernova originating from specific atomic transitions are compared with those of a laboratory spectrum. In this way the redshift $z \equiv (\lambda_0 - \lambda)/\lambda$ can be measured. (The subscript 0 stands for the measurement made by the observer on Earth.)

16. *How old is the universe and how is this learned?*

By looking deep into space one is looking back in time because of the time it takes light to travel to us. The look-back-time is related *approximately* to the Hubble constant and the redshift by $\Delta t \approx z/H$ (see Box 5). New data (2003) from measurements of radiation emitted before there were any stars show the universe to be about 13.7 billion years old, to within one percent (± 137 million years).

17. *Why is there only an approximate relationship for time, as in the above answer?*

Because times past cannot be directly measured. They depend on how fast the universe expanded, which has *varied* over time. But the amount by which light from a distant galaxy is redshifted depends uniquely on the amount by which the universe has expanded in the interim. And the redshift *is* measurable. So $z \equiv R_0/R - 1$ is an exact way of referring to the past (R denotes the scale of the universe at an earlier time and R_0 the scale now). For example, if the measured cosmological redshift of a distant object is $z = \frac{2}{3}$, then the scale of the universe was $R = \frac{3}{5}R_0$ of the present scale (size). If $z = 1.5$ then $R = \frac{2}{5}R_0$. What fraction of the present size was the universe for a group of galaxies at $z = 2$?

18. *Is R a measure of the size of the universe?*

It is a relative measure by which two eras of the universe can be compared by their respective scales.

19. *What is the largest presently measured redshift of an object?*

In 1994 the redshift of a galaxy 8C1435+635 was measured with an approximate value of 4.25. The universe then was $\frac{4}{17}$ of its present size. Two emission lines of ionized carbon and hydrogen were measured to obtain the red shift.

20. *How is the redshift of far-off galaxies measured?*

Light emitted by atoms at the periphery of stars in a galaxy can be detected by astronomers using telescopes: The light is emitted at unique energies or *wavelengths* corresponding to transitions of electrons between energy levels in the atoms. Each type of atom has its unique signature. The galaxy is receding because of the cosmic expansion, so the wavelength (λ) corresponding to each transition taking place in the receding galaxy is redshifted from that of a stationary atom (λ_0) in the observer's laboratory (i.e. $\lambda_0 > \lambda$). The fractional change in wavelength is directly related to the redshift also called the Doppler shift (see also Box 5) by $z = (\lambda_0 - \lambda)/\lambda$.

21. *What is a black hole?*

From General Relativity, it is found that if an (non-rotating) object of mass M lies within a sphere of radius $2MG/c^2$, where G is Newton's constant and c the speed of light, then it is a black hole — an object from which not even light can escape. (If the object is rotating, the relationship quoted is more complex.) Oppenheimer and Volkoff discovered the relationship in modern times. Though not precisely written down, it was discovered in 1783 by the Rev. John Michell and published in the *Proceedings of the Royal Society, Vol LXXIV*. The term "black hole" was not introduced until much after the theoretical investigation of these strange objects by J. R. Oppenheimer and H. Snyder in 1939. The name "black hole" was coined a few years later by John A. Wheeler, who earlier did not think they could exist — that some process would intervene to evade the collapse.

22. *Why do astronomers believe that a gigantic black hole occupies the center of the Milky Way?*

Near the center of the Milky Way, astronomers have observed stars that are moving in tight orbits at very high speeds (Figures 1.13 and 1.14). If they were not attracted toward the center by a very great mass, they would fly off into space. Using the measured velocities for these stars one can calculate, by Kepler's law, how much mass at the center is required to hold these visible stars. The massive object at the center of our galaxy has a mass of at least 300 million times the mass of our Sun, and its radius is too small to be other than a black hole.

Fig. 1.13 Image created in x-rays of the region of the Milky Way's central black hole which is called Sagittarius A*. The black hole itself cannot be seen; neither light nor anything else can escape it. Its location is near the center in the bright region of the multi-million degree gas, which is what is seen in this figure. Credit: Rainer Schödel (MPE) *et al.*, NAOS-CONICA, ESO.

23. *How could the radius of the central black hole in our galaxy be measured?*

Half of the angular separation of the two extremes (the approaching and receding stars) in Figure 1.14 and the distance from us to the Galaxy center provide R, the radius of the orbiting stars. The Doppler shift

2 Friedmann–Lemaître Equations

For the Robertson–Walker line element corresponding to a *homogeneous and isotropic* universe, only two of Einstein's (10 independent) field equations, $G_{\mu\nu} = -8\pi T_{\mu\nu} + \Lambda g_{\mu\nu}$, are independent [Narlikar (2002); Glendenning (2004)]. They can be taken as

$$\dot{R}^2 + kc^2 = (1/3)(\Lambda + 8\pi G\rho)R^2\,,$$

and

$$\ddot{R} = (1/3)[\Lambda - 4\pi G(\rho + 3p/c^2)]R\,.$$

Here, R is the scale of the universe in arbitrary units, $\rho = \epsilon/c^2$ is the sum of mass density of matter ρ_m and of radiation ρ_r, ϵ their total energy density, p the pressure, Λ Einstein's cosmological constant, and k is the curvature parameter.

Take the derivative of the first of the above pair, multiply the second by $\dot{R}$ and eliminate the Λ term from the resulting pair to obtain the conservation law implicit in the Einstein equations (divergenceless stress-energy tensor),

$$\dot{\rho} = -3(p/c^2 + \rho)(\dot{R}/R)\,.$$

This equation can also be written in two different ways:

$$d/dt(\rho c^2 R^3) = -p\, dR^3/dt$$

or

$$d\rho/dR = -3(p/c^2 + \rho)/R\,.$$

We can derive rigorously the behavior of radiation and matter densities as the universe expands. The equation of state for radiation is $p = (1/3)\rho_r c^2$. Therefore

$$d\rho/\rho = -4dR/R\,.$$

This yields the conservation equation for the equivalent mass density of *radiation*

$$\rho/\rho_0 = (R_0/R)^4, \text{ radiation.}$$

However, for *matter*, $p \ll \rho_m/c^2$, and we obtain instead

$$\rho/\rho_0 = (R_0/R)^3, \text{ matter.}$$

Thus radiation dominates early in the history of the universe, matter later, and finally after these have been diluted by the universal expansion, the cosmological constant dominates.

3 Einstein's Stationary Universe

At the time that Einstein applied his General Theory of Relativity to the universe, he, and most everyone else, thought that the world was stationary. Hubble had not yet discovered the expansion of the universe. To obtain a stationary solution, put $\dot{R} = \ddot{R} = 0$ into the two Friedmann equations (Box 2). We then obtain,

$$3kc^2/{R_0}^2 = \Lambda + 8\pi G\rho_0\,,$$

and

$$\Lambda = 4\pi G\rho_0\,.$$

Rearrange these to find,

$$\Lambda = kc^2/{R_0}^2.$$

Take the size of the universe to be its age, 13.5×10^9 years times the speed of light. Then, for $k = 1$,

$$\Lambda = 5 \times 10^{-36}(1/\mathrm{s}^2)\,,$$

and

$$\rho_0 = \Lambda/(4\pi G) = 6 \times 10^{-30}\ \mathrm{g/cm^3}\,.$$

Such a cosmological constant with such an average density would produce a stationary universe.

4 Redshift in Cosmology

In cosmology, a past event is usually referenced by the value of the redshift z because that is what can be measured, whereas the time at which it occurred cannot. An approximate time can be referenced only with an assumption of how the universe actually evolved with time, which of course is not measurable. The best one can do is to use a model of the expansion by reference to a particular scenario, say the Friedmann and Lemaître equation with definite assumptions about the cosmological parameters that appear in it. These parameters k, Λ, and the densities of matter and radiation ρ are known only within errors and there are three of them against one, the redshift z.

5 Redshift

The cosmological redshift of light emitted by a receding source, say a galaxy, is the fractional change in wavelength between that received by an observer λ_0 and that emitted by the source λ,

$$z \equiv (\lambda_0 - \lambda)/\lambda \geq 0\,.$$

Let R_0 denote the scale of the universe in the era of the observer, and R the scale of the universe at the earlier era when the light was emitted by the source. Then $R_0 > R$. (R is referred to as the scale factor of the universe, and represents size in relative units.) We note that in an expanding universe, wavelength of radiation is stretched in proportion to the expansion so that, $\lambda/\lambda_0 = R/R_0$. The very important connection between cosmological redshift z and the scale of the universe R follows from this:

$$z + 1 = R_0/R\,.$$

This tells us that for a distant object (seen as it was in the past because of light travel time) at a redshift of z, the scale of the universe now R_0, is larger than the scale R at that time in the past, by the factor $z + 1$. It follows that

$$\begin{aligned} z &= (R_0 - R)/R \equiv \Delta R/R = [(\Delta R/(\Delta t R)]\Delta t \\ &\approx (\dot{R}/R)\Delta t \equiv H\Delta t\,, \end{aligned}$$

where $\dot{R} \equiv dR/dt$. Here we have introduced the *Hubble constant*, $H \equiv \dot{R}/R$, (with the unit of inverse time) which expresses his discovery that the distant galaxies are receding from us at a velocity proportional to their distance,

$$v \equiv \dot{R} = HR\,.$$

The distance to the source (distant galaxy) is $R = c\Delta t$, where Δt is the time taken for light to travel from the source to the observer. Therefore, from above we have,

$$z \approx H\Delta t = (v/R)(R/c) = v/c \quad \text{(for small } z\text{)}\,.$$

This shows that the redshift of a not-to-distant galaxy is approximately equal to its velocity in units of light velocity if $v \ll c$.

6 Why Cosmological Redshift Rather than Doppler Shift

Note that the cosmological redshift cannot be properly called the Doppler shift. The latter refers to recession of an observed object in a fixed space, whereas the cosmological redshift refers to recession because space *itself* is expanding and the observed object is co-moving with the expansion. This is the case for the universe.

7 Three Epochs of the Universe

Einstein's equations describe, among other things, how gravity controls the expansion of the universe. For a uniform homogeneous universe they take the simple form, the Friedmann–Lemaître equation, which governs expansion and the continuity equation for energy conservation (see Box 2). During the radiation epoch (when the density of radiation dominated that of matter and the cosmological constant) the expansion equation takes the form;

$$\dot{R}^2 = 8\pi G\rho_0 R_0^4/3R^2$$

where ρ_0 and R_0 are the values of the density and scale factor at any convenient reference time (for example the present). We learn that the size of the visible universe increases with time in proportion to the square root of time, $R \sim \sqrt{t}$, but ever more slowly.

The mass densities of radiation and matter became equal at about a million years when the temperature was about 2000 degrees Kelvin; radiation slowly faded thereafter. When the matter density ρ_m is the dominant term of those on the right of the Friedmann–Lemaître equation, the universe entered the matter dominated epoch. The universal expansion is then controlled according to the equation

$$\dot{R}^2 = 8\pi G\rho_0 R_0^3/3R\,.$$

In this epoch, the expansion increases with time as $R \sim t^{2/3}$; the speed of expansion continues to decelerate because of the gravitational attraction of matter and radiation.

There is a *third epoch* when density has diluted and the dark energy term Λ dominates. The Friedmann–Lemaître equation becomes with time

$$\dot{R}^2 = \Lambda R^2/3\,.$$

The solution to the expansion equation for positive Λ, the dark energy, is $R \sim \exp\sqrt{\Lambda/3}\,t$. We can also note that $\ddot{R} = (\Lambda/3)R$, so that the universal expansion *accelerates* in the third era.

8 Redshift Correspondence with Time

The connection between redshift and time depends on the way the universe evolves. We do not know the time dependence of the expansion. The expansion proceeded differently according to the dominant contents of the universe. The Friedmann–Lemaître equations can be solved for the three epochs corresponding to radiation, matter and cosmological constant dominance. The connection is (where t_0 is the present time and t an earlier one, as measured from the beginning of the expansion):

$$1+z=\frac{R_0}{R}=\begin{cases}(t_0/t)^{1/2} & \text{radiation era}\\ (t_0/t)^{2/3} & \text{matter era}\\ \exp[\sqrt{\Lambda/3}\,(t_0-t)] & \text{dark energy era}\end{cases}$$

It is worth noting that the temperature of radiation falls as the universe expands because the wavelength of radiation stretches with the expansion;

$$\frac{T}{T_0}=\frac{R_0}{R}\,.$$

In the first two ages, the expansion decelerates because of gravity acting on the equivalent mass density of radiation at first and when that fades in importance, then on the mass. In the third, the cosmological term dominates and the cosmic expansion accelerates. The acceleration in the three ages is summarized as;

$$a\sim\begin{cases}-1/(t)^{3/2} & \text{radiation era}\\ -1/(t)^{4/3} & \text{matter era}\\ \Lambda/3\exp[\sqrt{\Lambda/3}\,t] & \text{dark energy era}\end{cases}$$

These connections can be derived as limiting cases from Friedmann's equation.

9 Present Baryon and Photon Density

From the present temperature of the radiation background, $T_0 = 2.728\,\mathrm{K}$, the *present* mass density of radiation can be calculated from the Steffan–Boltzmann law,

$$\rho_r(t_0) = aT_0^4/c^2 = 4.66 \times 10^{-34}\ \mathrm{g/cm^3}\,.$$

The present *number* density of photons (i.e. the number per cubic centimeter) can be found from Planck's law for black-body radiation and the measured cosmic background temperature. The number density is

$$n_\gamma(t_0) = 0.244(2\pi kT_0/hc)^3 = 413\ \mathrm{photons/cm^3}\,.$$

The baryon to photon ratio can be determined from primordial abundances (cf. [Glendenning (2004)], Section 5.4.4); it is $n_\mathrm{B}/n_\gamma = 5 \times 10^{-10}$. Hence, we can calculate the *present* baryon *number* density as

$$n_\mathrm{B} = 2 \times 10^{-7}/\mathrm{cm^3}$$

and the mass density is then found to be

$$\rho_\mathrm{B} = n_\mathrm{B} m_N = 3.5 \times 10^{-31}\ \mathrm{g/cm^3}\,.$$

As an added note we can emphasize that the number of photons vastly outnumbers the number of baryons,

$$n_\gamma = 2 \times 10^9\, n_\mathrm{B}.$$

As a result of this, photons effectively *did not* encounter baryonic matter or electrons from a very early time in the history of the universe.

Chapter 2

Formation of Galaxies

And God made two great lights, great for their use
To man, the greater to have rule by day, The less by night...

John Milton, *Paradise Lost, Seventh Book*

2.1. Fragmentation of Giant Clouds into Galaxies and Galaxy Clusters

The universe was born intensely hot, dense, and without even the seeds of the large-scale structures that we see today, such as galaxies. Until 100,000 years after this hot-dense beginning, matter was still almost uniformly distributed and the matter density was decreasing as the universe expanded. It took about 300,000 years to cool to a point where electrons could cease their frantic jostling in the heat of the universe to combine with nuclei to form neutral atoms of hydrogen and helium. Once the atoms are neutral, they no longer repel each other, and in the great clouds of these gases, even slight deviations about the mean value of the density acted as centers of gravitational attraction, drawing unto them surrounding matter, and in doing so, creating a void between them. These concentrations themselves became distinct smaller molecular clouds of hydrogen and helium with the mass of galaxies. Fragmentation continued into ever smaller and *denser* clouds of star-like masses. Thus *protostars*, still larger and less dense than stars, began to form.

The contrast between high- and low-density regions increased with time. By the time the Universe was a few hundred million years old, the densest regions had ceased to expand with the universal expansion, and began to collapse. While we cannot witness the process of galaxy formation, simulations using large supercomputers have been used to help visualize the process (Figure 2.1).

Galaxies formed in clusters — sometimes joined by filaments of galaxies.

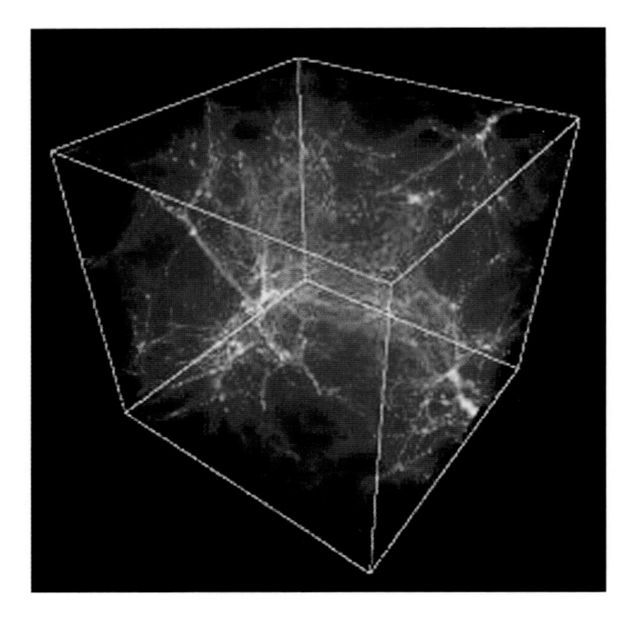

Fig. 2.1 This simulation of galaxy formation shows clusters of galaxies, the voids between them and filaments of galaxies connecting the clusters. The scale is enormous; filaments are (4 to 5) $\times$ 10^{21} kilometers long. What appear as cob webs comprise dots, each one representing a galaxy. Credit: Department of Astronomy, University of Texas at Austin, Paul Shapiro.

Interspersed among the galaxy-clusters are large regions devoid of galaxies. In the mathematical simulations of galaxy distributions, first knots of galaxies appeared in collapsing regions. Then filaments of collapsing regions spread out from these knots and after a time the filaments linked the knots together in a web of galaxies with knots, strings and voids (Figure 2.1). Besides matter, clusters of *galaxies* are believed to consist of approximately 10% of hot gas, and 2–5% of stars. The gases have temperatures of about 100 million degrees centigrade.

Ordinary matter (baryonic matter), such as the Earth, stars, white dwarfs, and neutron stars, makes up about 5% of the matter in the Universe. Of this, less than half is in visible stars.

About 85% of the matter in galaxies consists of an unknown type of matter called *dark matter*. Some of the dark matter may be ordinary (baryonic) matter consisting of white dwarfs, neutron stars and black holes. Neutron stars can be detected only in the neighborhood of our galaxy. Even of those that are in our neighborhood, most go undetected because the radio beam by which they might have been detected does not intersect our line of sight.[a] But by far the largest part of dark matter is non-baryonic, a form of matter whose nature is presently unknown to us.

The evidence of the existence of this dark matter is circumstantial; the formation of galactic structures — the knots, strings and voids — seems to have been unlikely without the help of unseen *dark matter*. Dark matter refers to matter and energy that cannot be seen, but whose gravitational effects on visible stars and galaxies reveals its presence. Fritz Zwicky, a Swiss-born astronomer at Caltech (in Pasadena, California) first suggested its existence in 1933. He observed that stars on the periphery of our own galaxy were moving too fast to be held by the gravitational grip of the observed stars in it. He also observed that our own galaxy and other nearby ones are moving toward a dense cluster of galaxies called the Coma Cluster. Moreover, he was able to determine — by the motion of galaxies on the periphery of the Coma Cluster — that it has about 400 times more mass than could be accounted for by the observable galaxies in the cluster. This became known as the missing mass problem, now known as dark matter.

2.1.1. Galaxies and Stars

Galaxies first formed in the densest spots in the Universe. At these locations the merging process started earliest and proceeded fastest, and they are now rich in elliptical galaxies. They are, in fact, the centers of what observers call rich clusters of galaxies: giant regions, perhaps ten million light-years in radius, within which gravity has successfully reversed the cosmic expansion. In these rich clusters, thousands of galaxies are moving at speeds up to 1000 kilometers per second through a diffuse sea of dark matter, gas at temperatures of about 10,000,000 degrees centigrade, and stars that have escaped from individual galaxies.

The degree of compression in forming a star from the initial nebulous clouds is enormous as is known from observations made on present galaxy and star formation in distant galaxies.[b] Typically the change in density from the collapsing tenuous cloud to a star is 20 orders of magnitude. The gravitational energy that is gained by their collapse is converted into heat energy. With increasing density, the clouds become opaque to the loss of radiation that would otherwise carry some of the energy away. Finally,

[a]The estimate can be made on the basis of average beam width and radio telescope efficiency.

[b]Distant galaxies, before they were recognized as such, were called *quasars*, meaning quasi-stellar objects.

Fig. 2.2 A deep view into space beyond the Milky Way to myriad distant galaxies of various types. Though the field is a very small sample of the heavens, it is considered representative of the typical distribution of galaxies in space, because the universe looks largely the same in all directions (referred to as the homogeneity and isotropy of the universe). Credit: Hubble Space Telescope; Participating institutions: MPA für Astrophysik Garching (D), Sterrewacht Leiden (NL), Univ. Cambridge (GB), Institut d'Astrophysique de Paris (F), Univ. Durham (GB), Osservatorio Astronomico di Padova (Italy), NASA, ESA, Beckwith (STSci) and the HUDF Team.

as the collapse continues, the heat becomes so intense that thermonuclear reactions are ignited; they convert the light elements — hydrogen and helium — into heavier elements, carbon, and oxygen up to iron. For tens of millions to billions of years, depending on the mass of the star, these reactions create heat and radiation, whose pressure resists collapse in luminous stars like our Sun. Our Sun is about 4.5 billion years old and will live for another 7 billion years or so as a luminous star before it fades away to leave behind a small white dwarf surrounded by a beautiful planetary nebula as in Figure 4.1.

Beyond our own Milky Way galaxy, which contains an estimated 200–400 billion stars, the universe is populated by countless other galaxies of various forms and sizes as the Hubble space telescope shows us by its deep view into space (Figure 2.2).

2.2. Types of Galaxies

Galaxies come in various forms, which were classified by Edwin Hubble. His classification, shown in Figure 2.3, is still used. We recount some of their main characteristics in the following.

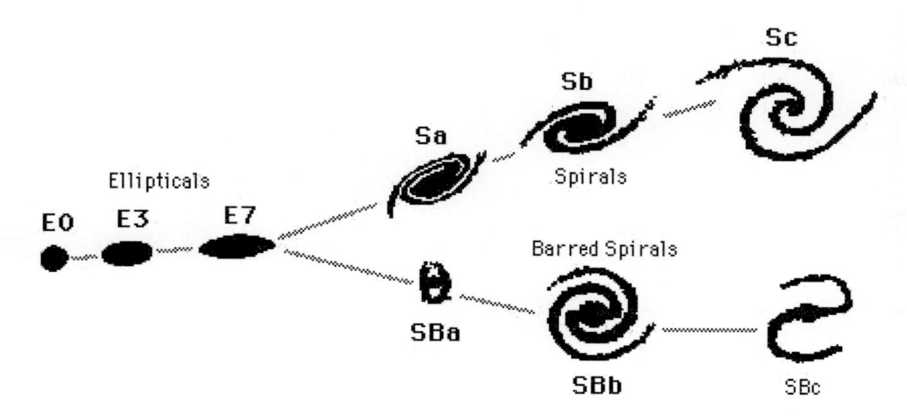

Fig. 2.3 Types of galaxies as classified by Hubble which separates most galaxies into elliptical, normal spiral, and barred spiral categories; sub-classifications are according to properties such as the amount of flattening for elliptical galaxies and the nature of the arms for spiral galaxies. The galaxies that do not fit these categories are classified separately as irregular galaxies.

2.2.1. Spiral Arm Galaxies

A spiral is very natural form of galaxy due to the way they are formed by collapsing rotating gases, which later subdivided to form stars. Most galaxies are spirals, like the Andromeda Galaxy seen in Figure 1.1. It is several times more massive than the Milky Way as measured in luminous stars, but is otherwise similar in form. The Milky Way mass is $\approx$ 1000 billion sun masses ($M_{\text{Milky Way}} \approx 10^{12} M_{\odot}$). Barred galaxies (Figure 2.4) are extreme forms of spiral galaxies that do not have enough mass at their centers to maintain the many arms of a spiral.

Fig. 2.4 A barred galaxy NGC 1365, is a supergiant galaxy with a diameter of about 200,000 light-years. It is moving away at 1632 km/sec. One full turn of the bar will take about 350 million years. Credit: Volker Wendel and Bernd Flach-Wilken.

The youngest stars in a spiral galaxy reside in the thin disk, older stars in the central bulge, and the oldest, in *globular* clusters that have highly elliptical orbits of huge dimensions that take them far outside the central bulge of the galaxy, even to distances greater than the diameter of the entire disk.

A close face-on view of the beautiful galaxy called the Whirlpool (M74) is shown in Figure 2.5. It is having a close encounter with a nearby companion galaxy off the upper edge of this image. The companion's gravitational pull is triggering star formation in the main galaxy, as seen in brilliant blue detail by numerous, luminous clusters of young and energetic stars.

Aside from mythological names, like Andromeda (Princess of Ethiopia), or descriptive names, like the Whirlpool, astronomers also give them a number as their discovery is recorded in a catalogue. A French astronomer began a very early catalogue; Charles Messier (1730–1817) catalogued 109 *bright* nebulae, including as the first, the Crab Nebula (M1). Stars in that catalogue

Fig. 2.5 A close face-on view of a beautiful spiral galaxy called the *Whirlpool*, M74, about 30 million light-years away toward the constellation Pisces. The red blotches in this composite view are *ultra luminous x-ray sources* (ULX) mapped by the Chandra X-ray Observatory. They radiate 10 to 1,000 times more x-ray power than ordinary x-ray binary stars, which harbor a neutron star or stellar mass black hole. Their x-ray brightness changes in periods of 2 hours or so; consequently, astronomers conclude that they may be intermediate mass black holes — black holes with masses 10,000 times or so greater than the Sun, but still much less than the million solar mass black holes which lurk in the centers of large spiral galaxies. Credit: NASA and the Hubble Heritage Team (STScI/AURA).

are referred to by a sequence number following M, including Andromeda (M31) in Figure 1.1 and the Whirlpool (M74) in Figure 2.5. William Herschel (1738–1822), his sister Caroline, and later William's son, John, began a systematic cataloguing of all nebula visible through various telescopes of William's own making. This eventually resulted in the *New General Catalogue* (NGC) containing several thousand entries. William (Figure 2.6), originally a musician, was a self-taught astronomer who eventually built what was then and remained for many years, the world's largest (40-inch

Fig. 2.6 William Herschel as a young man, a musician and self-taught astronomer who eventually was appointed Court Astronomer by King George III. He built the largest reflecting telescope of its time.

reflecting) telescope. He was recognized by King George III of England as the Court Astronomer with an annual salary and during the course of his astronomical career, discovered about 2500 nebula and star clusters. His sister too, Caroline, was recognized by the crown with an annual salary and became famous in her own right, being honored by many of the crowned heads of Europe.

2.2.2. Elliptical Galaxies

Aside from the initial blobs of stars formed when collapsing molecular clouds first condensed to form galaxies, a galaxy does not naturally form an elliptical shape. Rather, two spiral galaxies that collide because of their mutual gravitational attraction, may pass through each other, disrupting their shapes but otherwise leaving the stars intact. However, on subsequent collisions, as their kinetic energy of motion is dissipated by the collisions, a galaxy merger may result. This generally leads to a blob or elliptical galaxy (Figure 2.7). An elliptical galaxy is fairly diffuse at its edge but rises very sharply in den-

Fig. 2.7 An elliptical galaxy with a super massive black hole at its center. The core of the galaxy, M87, is the bright dot (center right) but the galaxy includes the fainter halo of stars which extends at least 100 core diameters beyond. The black hole has ingested stars equal to 2 billion times our Sun's mass. And this cataclysmic activity is casting out a jet of high energy electrons and radiation extending toward the left of the frame. The length of the jet is 5,000 light-years at optical wavelengths and 100,000 light-years at radio wavelengths. Image credit: NASA and the Hubble Heritage Team (STScI/AURA). Acknowledgments: J. A. Biretta, W. B. Sparks, F. D. Macchetto, E. S. Perlman (STScI).

sity toward its center and usually harbors a giant black hole. Nearby stars, attracted by the black hole's intense gravity, spiral inward, destroying each other as they crowd around it just before being ingested.

2.2.3. Lenticular Galaxy

NGC 5866 is a beautiful lenticular galaxy seen almost edge-on, Figure 2.8. The dark dust lane is tilted by about 2 degrees against the galaxy's symmetry plane. This galaxy is the brightest of a remarkable group of galaxies, lying 40 to 50 million light-years distant. From the dynamics of a group of nearby

Fig. 2.8 A beautiful lenticular (like a lens) galaxy, NGC 5866. Its mass is estimated to be 1 trillion solar masses [E.M. and G.R. Burbidge (1960)]. Photo credit: Bernd Koch and Stefan Korth, near Solingen, Germany.

galaxies, the husband and wife team of E. M. and G. R. Burbidge (1960) have estimated the mass of NGC 5866 to be about one trillion solar masses (10^9 sun masses); it is moving through space at a velocity of about 900 kilometers per second. Probably the first astronomers to see it were the remarkable brother and sister William and Caroline Herschel, when William determined its position on May 5, 1788.

2.3. Fritz Zwicky and Galaxies as Gravitational Lenses

Optical lenses are familiar to us, but it may be surprising that the gravity of large galaxies or clusters of galaxies can act like a *gravitational* lens. It was Fritz Zwicky who conceived this remarkable notion, and George Abell was the prime motivator for the construction of a special wide-field telescope at Mount Palomar in Southern California with which the effect could be discovered. It was designed for a search of galaxy clusters. The rich galaxy cluster known as Abell 2218, is so massive and compact that light rays

Fig. 2.9 A spectacular example of gravitational lensing, predicted by Fritz Zwicky, can be seen as the arc-like pattern spread across the picture. It is an illusion caused by the gravitational field of the cluster of foreground galaxies called CL 0024+1654. This is a large cluster of galaxies located 5 billion light-years from Earth. It is distinctive because of its richness (large number of member galaxies), and its magnificent gravitational lens. The blue loops in the foreground are lensed images of a spiral galaxy located behind the cluster. In one such case the lensed object was 50 times fainter than objects that can be seen with ground-based telescopes (Andrew Fruchter). The image above was taken by Colley, Tyson and Turner using the Hubble Space Telescope. The distance of the lensed object behind the foreground cluster is presently unknown. Credit: HST and NASA.

passing through it are deflected by its enormous gravitational field, much as an optical lens bends light to form an image (Figure 2.9). However, unlike optical lenses, which are ground to high precision, the galaxy masses and spacing in the cluster are not nearly so regular as to produce a high definition image. Very distant galaxies, which are 50 times fainter than objects that can be seen with ground-based telescopes, show up as arcs. Distant galaxy populations can be as much as 5–10 times farther than the lensing cluster itself. Distant populations can be seen as they were when the universe was just one-quarter of its present age.

Fritz Zwicky (1898–1974), born in Hungary, was an eccentric Swiss-American astronomer. He was professor of astronomy at Caltech.

Fig. 2.10 Fritz Zwicky, an eccentric Swiss-American astronomer, was the first to understand that large collections of stars could act as a *gravitational lens*, bringing into evidence stars that are far beyond them and out of range of our telescopes. He also had the fertile imagination and foresight to predict that supernova, the explosion of very large stars at the end of their lives, would leave behind a neutron star or black hole, long before either had been discovered.

Zwicky had a difficult personality, and intentionally intimidated practically everyone he encountered. Zwicky (Figure 2.10) was also fond of calling people spherical bastards because they were bastards every way he looked at them. He was known to accost unfamiliar students in the astronomy building at Caltech with the interrogation "Who the hell are you!?" and he is reputed to have remarked to the head of his department at Caltech "I have a good idea every two years. Give me a topic, I will give you the idea!" He usually lived up to his self-assessment.

Zwicky had a vivid imagination; some of his contemporaries considered it wild. Besides being the first to conceive the possibility that clusters of galaxies could act as a gravitational lens bringing into view galaxies so distant that they could not be seen by man-made telescopes (Figure 2.9), he also proposed the amazing notion of *dark matter* to account for the high velocities of stars at the periphery of our own and other galaxies that would fly off into space if gravitationally bound only to the visible stars. He too foretold, in collaboration with Walter Baade, that a supernova, the explosion of a large star at the end of its life, might leave behind a neutron star, as yet unknown. It was not until many years afterward that the first such star, quite by accident, was discovered by a graduate student, Jocelyn Bell, at Cambridge University (see page 79).

2.4. Unseen Matter and Energy

2.4.1. Dark Matter, Dark Energy

One of the great mysteries concerning our universe, and even our own galaxy, is that we know that it contains much more matter than can be accounted for in visible stars. Fritz Zwicky already suspected this in the 1930s. From the motion of stars near the boundary of the Milky Way, and also from the movement of the Milky Way in relation to a nearby galaxy cluster, he could tell that there was much more matter in that galaxy that exerted a gravitational attraction on ours than could be seen in stars. In fact, we know also from the synthesis of helium in the first few minutes what the density of matter is, and it amounts to only about 5% of what can be seen by astronomers in stars. The *missing mass* is referred to as dark (baryonic) matter. Presumably it resides in white and brown dwarfs, too distant or faint to see, as well as in neutron stars, black holes, and distant planets.

But there is a greater mystery! From the behavior of the cosmic expansion as a whole it has become evident from observations made in the past few years that there is a completely different kind of energy called *dark energy* that is driving the universe into an *accelerating* expansion. That the universe is expanding has long since been known from Hubble's famous discovery. That the expansion is actually accelerating has been discovered in our own day. The agent driving the acceleration has been called *dark* energy; while its effect is known, its nature is not. So the story of dark energy has not ended, but it had its beginning many years before the actual discovery of the effect, in the work of Albert Einstein.

But in Einstein's day, the cosmic expansion had not yet been discovered. He believed as Newton before him, that the universe was static. It was a fixed belief in Western thought that the world continued the same from everlasting to everlasting. Realizing that his equations would lead either to contraction or expansion, he introduced the cosmological constant into his theory of gravity to exactly counterbalance the pull of gravity on the universe as a whole, which by itself would slow the expansion and perhaps cause a re-collapse. Only after Hubble's discovery of expansion, did Einstein realize that he could have predicted the cosmic expansion.

However, even before Hubble's discovery of expansion, Alexandre Friedmann, the Russian meteorologist and bomber pilot, and Georges Lemaître, the Belgian priest turned cosmologist, understood Einstein's theory, and with it, independently described the expansion of the universe as the consequence of a dense and hot beginning. They also foresaw, and described mathematically, the possibility that the universe might expand forever, even at an accelerating rate. Seventy years later Saul Perlmutter — leading a large team of researchers at the Lawrence Berkeley National Laboratory — did indeed discover exactly that — accelerating cosmic expansion. And it is the sign and magnitude of the cosmological constant that represents the present, and presumably everlasting acceleration. So many years after Einstein had introduced the cosmological constant into his theory, and then called it his greatest mistake, it has returned; his cosmological constant represents what is now called the constant *dark energy* that pervades the expanding universe, driving it now into acceleration.

2.4.2. Hot and Cold Dark Matter

Hot dark matter refers to particles that have zero or near-zero mass, such as the neutrino. The nature of the world we live in — as Einstein and Minkowski discovered — requires that massless particles move at the speed of light. Thus, such very low-mass particles must move at very high velocities and therefore form very hot gases.

On the other hand, *cold dark matter* is composed of objects sufficiently massive that they move at sub-relativistic velocities, and therefore form much colder gases. The difference between cold dark matter and hot dark matter is significant in the formation of structure because the high velocities of hot dark matter causes it to wipe out structure on small scales, whereas cold dark matter aids in the formation of galaxies through gravity.

2.4.3. Distinguishing between Dark and Ordinary Matter

Ordinary matter (baryonic matter) such as the Earth, stars, white dwarfs, and neutron stars makes up about 5% of the matter in the Universe. Of this, less than half is in visible stars. Part of the non-baryonic dark matter may be *hot dark matter* consisting of neutrinos.[c] The nature of the other part of the hot dark matter is unknown. It differs from ordinary matter

[c]The neutrino is almost, but not quite massless. The most common type (electron neutrino) has a mass of about 10 eV or about one-hundred-millionth of the proton mass, $10^{-8}m_p$.

in being able to pass through ordinary matter. It is assumed to consist of weakly interacting particles (called WIMPS) of a type as yet undiscovered; all attempts to detect them in the laboratory have failed.[d] But hot dark matter is believed to reside now mainly in the halos of galaxies, as Fritz Zwicky already surmised in the 1930s from the motion of stars near the boundary of the *visible* part of the Coma cluster.

2.5. Globular Clusters

Globular clusters are tightly knit bunches of stars of approximately ten thousand to one million stars. They populate the halo or bulge of the Milky Way and of other galaxies with a significant concentration toward the Galactic Center. Globular clusters make enormous excursions out of the central halo.

Of the over 200 globular star clusters that orbit the center of our Milky Way galaxy, 47 Tucanae is the second brightest (behind Omega Centauri). Known more briefly as 47 Tuc or NGC 104, it is only visible from the Southern Hemisphere. Light takes about 20,000 years to reach us from 47 Tuc which can be observed near the Small Magellanic Cloud in the constellation of Tucana. Red giant stars are particularly easy to see in Figure 2.11. The dynamics of stars near the center of 47 Tuc are not well understood, particularly why there are so few binary systems there.

Generally, globular clusters abound with neutron stars which, however, are not optically visible; rather they are detected by large radio telescopes. Because of their high population of remnant neutron stars, and also the fact, learned from spectroscopic studies, that stars in globular clusters have a smaller proportion of heavy elements than the stars in the disk of the Milky Way Galaxy itself, globular clusters must be very old.

Stars that have a small iron content are recognized as being very old. The reason is that the universe started its life with no iron: only light elements — hydrogen, helium and a little lithium — were made in the first few minutes. Heavier elements were made much later during the life of stars. At their death in a supernova explosion, their contents are distributed into the universe; there they wander for eons before being gathered together into great clouds of which portions begin a gravitational collapse. Galaxies of stars are formed during that time. As each generation of stars lives and dies, the universe is further enriched in heavy elements.

[d]Recent experiments by physicists from a consortium of universities tentatively have claimed the discovery of the weakly interacting particle and put a tentative value on its mass that is 60 times heavier than the neutron.

Fig. 2.11 47 Tucanae is the second brightest globular cluster in our Milky Way galaxy. The stars of 47 Tucanae are spread over a volume nearly 120 light-years across. Besides luminous stars, globular clusters are rich in neutron stars and white dwarfs. To have such a population of mature stars indicates that globular clusters are very old. The globular cluster 47 Tucanae is approaching the solar system at roughly 19 km/s. Red giant stars are also easily seen. This view shows only the edge of the cluster. Credit: D. Hunter (Lowell Observatory, STScI) *et al.*, HST, NASA.

2.6. Box 10

10 Galaxy Cluster Data

A parsec is an astronomer's unit of distance: A star at a distance of 1 parsec shows an annual parallax (caused by Earth orbiting Sun) of 1 second of arc.

1 parsec = 3.1×10^{13} km = 1.9×10^{13} miles = 3.3 light-years
1 million parsecs (Mpc) = 3.1×10^{19} km = 1.9×10^{19} miles = 3.3×10^{6} light-years

Filaments: 140 Mpc = 4.5×10^{21} km
Supercluster: 70 Mpc = 2.2×10^{21} km
Galaxy cloud cluster: 20 Mpc = 6.4×10^{20} km
Milky Way diameter: 100,000 light-years = 9.5×10^{17} km

Chapter 3

Birth and Life of Stars

A star is drawing on some vast reserve of energy .... This reservoir can scarcely be other than the subatomic energy .... There is sufficient in the Sun to maintain its output of heat for 15 billion years.

Sir Arthur Eddington (1920)

3.1. The First Stars

The lightest elements were made in the first few minutes in the life of the universe. The universe expanded and cooled too quickly to continue the synthesis of heavier ones. Elements heavier than lithium and beryllium, up to the mass of iron, can be made only in stars through the slow process of fusing lighter elements to make heavier ones. A small amount of energy in the form of heat is released in the process. These fusion reactions, called thermonuclear fusion, are the processes by which stars create heat and pressure. The pressure resists gravity and sustains stars like our Sun for billions of years as they slowly collapse.

But the first stars, which formed as early as 200,000 years after the beginning, were giants as compared to our Sun — as much as 100 to 1000 times its mass. From their flaming surfaces, molecular dust was cast off, which formed the nuclei around which new stars like our Sun condensed. Meanwhile, gravity, acting on the enormous mass of the giants, crushed them soon after birth; they were driven rapidly through their life cycle of burning the primeval hydrogen to form helium, and from these helium nuclei to form carbon and oxygen and ever heavier elements up to iron. These elements formed shells around a spherical core of iron at the center of the star. As the iron core grew to a critical mass — called the *Chandrasekhar mass limit* after the great Indian astrophysicist and longtime professor at the University of Chicago — gravitational attraction became so strong that the core suddenly collapsed from an object the size of 1000 kilometers to one of 15.

It took but a small fraction of a second to do so. The core then erupted in a supernova explosion. Thus were the first atoms — so necessary for planets to form and life to emerge — dispersed into the universe. The same sequence of processes occur to this day in later generations of stars, including our Sun, except that our Sun and other low-mass stars live longer, produce only very light elements, and die a gentler death (see page 102).

3.2. Star Birth

In the vast regions of space between the stars there are enormous clouds of molecular gas, composed mostly of the primeval hydrogen and helium elements. The masses of these clouds are up to several hundred thousand Sun masses and they have typical diameters of 50 light-years (500,000 billion kilometers); the gas pressure in such clouds is sufficient to balance the effects of gravity and they remain quasi-stable for long periods. But the reverberation of a distant supernova, the death and explosion of a mature star, can act as a trigger for the collapse of a *part* of the cloud by causing a wave-like motion of alternating regions of denser and rarer concentrations. The collapse may take half a million years or more, ultimately forming stars. Sir James Jeans, in 1902, first understood this process, and what is more, derived a way to calculate the mass necessary for collapse under the changing conditions of the evolving universe.

The Eagle Nebula is one such spectacular location of starbirth. In Figure 3.1 we see *into* a large shell of dust, in which a cluster of stars is being born. The Eagle is an emission nebula — meaning that the stars within provide its light. It lies about 6,500 light-years away, spans about 20 light-years, and is visible with binoculars toward the constellation of Serpens.

3.3. Orion: A Stellar Nursery

The Orion Nebula is an immense molecular cloud of hydrogen and helium gases. It glows from the light of hot young stars — a stellar nursery — only 1,500 light-years away. This is very close to us considering that the Milky Way galaxy is about 90,000 light-years across. The Great Nebula in Orion can be found with the unaided eye in the constellation Orion. In the view shown in Figure 3.2 the bright stars of the Trapezium in Orion's heart are seen together with sweeping lanes of dark dust that cross the center; the red

Fig. 3.1 The bright region of the Eagle Nebula (M16) provides a window into the center of a larger dark shell of dust. Through this window, a brightly lit region appears where a cluster of stars is being formed. Already visible are several young bright blue stars whose light and winds are burning away and pushing back the surrounding filaments and walls of gas and dust. The Eagle Nebula lies about 6,500 light-years away and spans about 20 light-years. Credit and copyright: T. A. Rector and B. A. Wolpa, NOAO, AURA, NSF.

glowing hydrogen gas, and the blue tinted dust reflect the light of newborn stars. The Orion Nebula, which includes the smaller Horsehead Nebula, will slowly disperse over the next 100,000 years.

Fig. 3.2 A stellar nursery known as the Orion Nebula whose glowing gas surrounds a nursery of hot young stars at the edge of an immense interstellar molecular cloud only 1500 light-years away. Credit and copyright: Robert Gendler.

3.4. Birth of Massive Stars in the Trifid Nebula

We are familiar with optical telescopes that magnify images that we can see with our eyes. Our eyes are sensitive to an extremely small band of wavelengths in what is called electromagnetic radiation. X-rays form another part of such radiation having a much smaller wavelength compared to light, while infrared (heat) has very long wavelengths and radio waves even longer.

Fig. 3.3 The Trifid Nebula is a giant star-forming cloud of gas and dust located 5,400 light-years away in the constellation Sagittarius. The earliest stages of formation of *massive* stars are shown in this figure. Credit: Gay Hill and Whitney Clavin, Jet Propulsion Laboratory, Pasadena, California, Dr. Giovanni Fazio, Smithsonian Astrophysical Observatory, Cambridge, Mass. and NASA's Jet Propulsion Laboratory, Pasadena, California.

Astronomers use all parts of the spectrum with appropriate instruments for their detection. The wavelength of visible light, centered on yellow, is a little less than a millionth of a meter. The striking image seen in Figure 3.3 taken with an infrared telescope shows a vibrant cloud called the Trifid Nebula dotted with glowing stellar incubators. Tucked deep inside these incubators are rapidly growing embryonic stars, whose warmth NASA's Spitzer Space Telescope was able to see with its powerful heat-seeking eyes. This view offers a rare glimpse at the earliest stages of massive star formation — a time when developing stars are about to burst into existence. Massive stars develop in very dark regions so quickly that it is difficult for astronomers to find them in the act of forming.

The Trifid Nebula (Figure 3.3) is located 5,400 light-years away in the constellation Sagittarius. Previous images, taken by the Institute for Radio Astronomy Millimeter Telescope in Spain, revealed that the nebula contains four cold knots, or cores, of dust, which are *incubators* where stars are being born. Astronomers had thought that the incubators in the Trifid Nebula were not yet ripe for star birth. But, the Spitzer telescope, with infrared eyes sensitive to hot spots, found that they had already begun to develop warm stellar embryos.

The Trifid Nebula is unique in that it is now dominated by one massive central star, a mere 300,000 years old. Radiation and winds emanating from that young hot star have sculpted the Trifid cloud into its current cavernous shape. These winds have also acted like shock waves to compress gas and dust into dark cores, whose gravity caused more material to fall inward until embryonic stars were formed. In time, the growing embryos will accumulate enough mass to ignite and explode out of their cores like baby birds busting out of their eggs.

Compared to the age of our Sun, 4.5 billion years, the giant star in the Trifid Nebula is a mere youth — very hot and turbulent. It is fortunate for us and all life on Earth, that the violent years of our Sun's youth are far behind.

3.5. Galaxy Collisions and Star Birth

It may at first seem surprising that galaxies sometimes collide given the vastness of space. But there is also the vastness of time. Galaxy collisions actually stimulate the birth of stars. A spectacular example of the result of a collision is seen in the Cartwheel Galaxy (Figure 3.4). A small intruder galaxy at some time in the past has careened through the core of another galaxy, sending a ripple of energy into space, which is plowing gas and dust before it. "This cosmic tsunami, expanding at an estimated 200,000 miles per hour, leaves in its wake a firestorm of newly created stars. The bright blue knots are gigantic clusters of newborn stars. Immense loops and bubbles have been blown into space by exploding stars (called supernovae), going off like a string of firecrackers." [a]

Though infrequent, the result of another spectacular collision of galaxies is shown in Figure 3.5. The gravitational shock caused by the collision drastically changes the orbits of stars and gas in the galaxy's disk, causing them to rush outward, like ripples in a pond after a rock has been thrown in. As the ring of stars and clouds plow outward, gas clouds collide and are compressed to the point that they contract under their own gravity, collapse, and form an abundance of new stars. These newly formed stars, which are

[a]Kirk Borne, STScI.

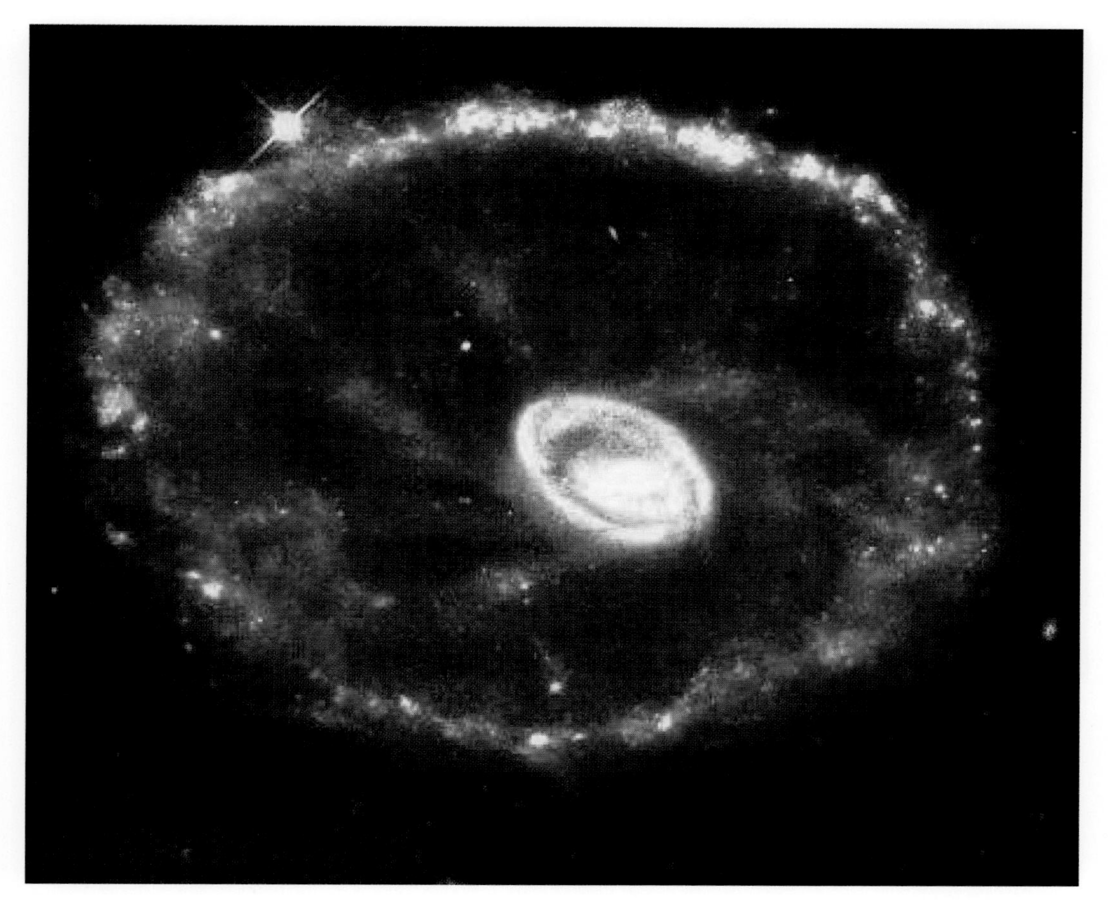

Fig. 3.4 A smaller galaxy — careened through the core of a galaxy, called the Cartwheel, disrupting it and lighting a ringed fire of star formation and of supernovae. Credit: Kirk Borne (STScI), and NASA.

blue, account for the color of the ring. Creation of stars in a group is usual; our Sun — a solitary star — is a rare exception to the rule.

3.6. In the Fires of Stars

The very lightest elements were forged in the intense fires of the first few minutes. Two-thirds of the atoms in our bodies were made in those few minutes — fourteen billion years ago. After several hundred seconds had passed, the universe — because of its expansion — had cooled too much for further building of a heavier element by combining several lighter ones.

Fig. 3.5 A ring of brilliant blue *clusters* of stars wraps around the yellowish nucleus of what was once a normal spiral galaxy like the Milky Way. A close encounter with another galaxy caused its distorted shape. Typical galaxies such as the Milky Way contain 400 billion stars and have a mass of 10 billion Suns. Credit: NASA, ESA, and the Hubble Heritage Team (AURA/STScI), J. Higdon (Cornell U.) and I. Jordan (STScI).

Elements heavier than lithium and up to iron were made much later after the first stars were born; still heavier elements were made at the peripheries of those stars at the end of their lives in supernova explosions.

The processes by which heavier elements are built from lighter ones in the hot interior of stars are called thermonuclear fusion reactions. When, for example, two carbon nuclei are fused to form magnesium, the mass of the latter is less than the two carbons. Conservation of energy is an immutable law of nature, and Einstein discovered that there is an equivalence between mass and energy according to his much quoted law, $E = mc^2$. The mass-energy appears as heat-energy. Such are the processes that convert some mass to heat in stars in the amount mc^2. And it is the pressure exerted by the heat energy that resists the pull of gravity, which would otherwise cause the star to collapse immediately.

Indeed, collapse is inevitable, but delayed for millions to billions of years by the internal fires. The Sun will live for about 12 billion years — lighter stars, far longer — heavier stars much shorter. Gravity's grasp on stars ten times larger than our Sun will speed them to their explosive end in a few million years, and in the case of very massive ones — fifty or more times massive than our Sun — envelop their cores in a black hole.

The very earliest stars were indeed massive, lived a short time, and were composed only of the primordial hydrogen and helium. Nevertheless, successive generations of these rapidly evolving stars produced a large fraction of the heavy elements that exist in the universe today.

As Joseph Silk [Silk (2001)] puts it, "we are children of the stars." A large fraction of the heavier elements were produced only 200,000 years after the beginning in the first giant stars. And two-thirds of the atoms in our bodies were made in the first few minutes. These atoms wandered for eons in great clouds, which eventually fragmented under gravity's attraction into galaxies, then stars and planets.

3.7. Thermonuclear Evolution of Stars

The first element made was hydrogen: it is the predominant element in the universe and in a young star. But gravity squeezes the matter of a newly forming star making its interior very dense and hot. Heat is energy, and the energy is in the form of motion — the hotter — the more violent the motion. In this hot environment, the electrons of the atoms are torn away laying bare the electric charge of the hydrogen nuclei. They therefore repel each other but in their violent motion the repulsion is overcome and four hydrogen fuse to form helium. The helium weighs less than the four hydrogen and the excess energy appears as heat — the kinetic energy of motion which exerts pressure. This source of pressure resists the pull of gravity and sustains the star against collapse as long as there is nuclear fuel to burn (as in the fusion of four hydrogen to form helium) and temperatures are high enough to burn it. When a small star like our Sun has consumed its hydrogen it will settle fairly quietly into a white dwarf, about 1/100 the original size.

In a star five times the mass of the Sun or more, a whole sequence of nuclear burnings take place — three helium to make carbon, and so on — the heavier elements sinking toward the center where it is hotter and new cycles of combustion take place until an onion-ring like structure of elements is formed, the heaviest at the center (Figure 3.6). At each stage in this process, the star contracts a little. Iron is the heaviest element produced *inside* a massive star. Fusing elements to form elements heavier than iron requires an *input* of energy rather than producing the energy that is necessary

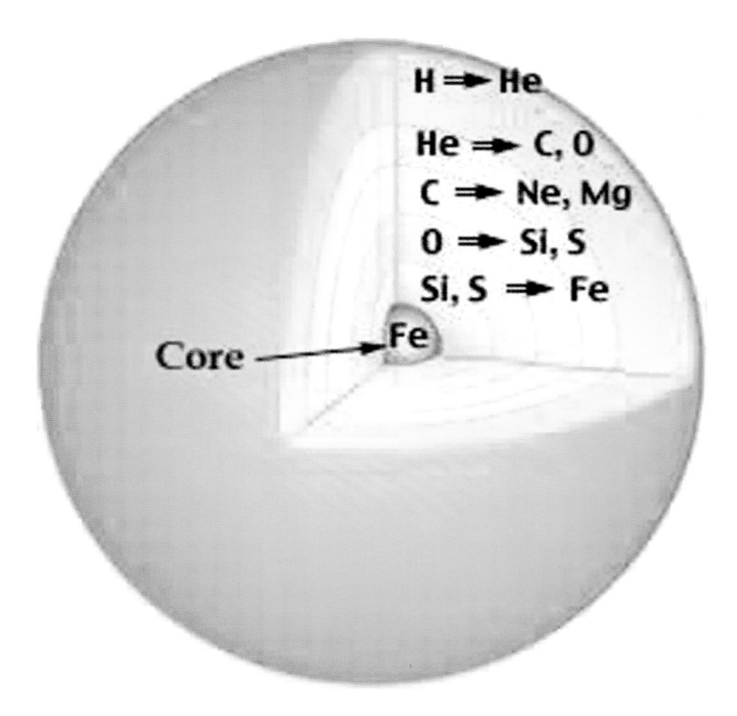

Fig. 3.6 A mature star has undergone a series of thermonuclear reactions forging the heavier from the lighter ones. Concentric shells of ever heavier elements are formed as the heavier elements sink to the center. It is ever hotter toward the center, fusing hydrogen to helium near the surface, and heavier elements in the interior. When the mass of the iron core grows to a value equal to the Chandrasekhar limit the star will collapse from an object about the size of Earth to a ten-mile object in a second. As a result a tremendous explosion occurs — a supernova explosion — producing a neutron star or black hole and hurling the remains of the original star into space to form a nebula like the Crab in Figure 4.3. The Crab's progenitor star had a mass of about 9 Sun masses with such an interior of onion-like layers. Credit: NASA.

to sustain the star's burning processes and create the heat and pressure that sustain it against gravity's grip and the eventual collapse of the star. When the iron core mass reaches a critical value, called the Chandrasekhar limit (page 77), the core collapses, then bounces, creating a supernova explosion from the gravitational energy acquired during the collapse. What is left is a remnant of much less mass — but very dense — a *neutron star*.

The collapsed core of a star that is 50 or more times the mass of the Sun, also ends its life in a supernova, but leaves behind a black hole having a few times the Sun's mass. The remainder of the energy of the original star's mass is accounted for by the explosive energy and, by far the largest part, an immense shower of ghostly neutrinos traveling at almost the speed of light.

3.8. Questions

1. *What is the evolutionary scenario in a massive star's evolution, that leads to a supernova explosion?*

   During most of its life, in the high-pressure core, gravitational energy is converted into heat energy, which drives the thermonuclear fusion processes by which the primordial hydrogen is combined to form helium, and from this, 3 helium nuclei into carbon, and so on to iron. The heavier elements sink to the center. This chain of reactions ceases with the production of iron, because up to that point, each fusion reaction *produces* energy in the form of heat, whose pressure which resists gravity and sustains the star against collapse. Iron is the most strongly bound nucleus so energy is required to fuse heavier elements beyond iron. Consequently, the processes that led to an output of heat pressure ceased. Iron is therefore the heaviest element that can be produced *inside* stars. When sufficient iron has been produced in this fashion, namely the Chandrasekhar mass of about 1.4 Sun masses, the internal pressure in the iron core can no longer resist gravity, and the core collapses in a small fraction of a second, destroying all nuclei and rendering them into their constituents, the neutrons and protons. The collapse of the core of nucleons continues until they are squeezed so tightly that the nuclear force — which is attractive at the normal inter-nucleon distance in nuclei but is repulsive at very short distance — causes them to repel each other and causes the core to bounce. This initiates the supernova explosion of the star.

2. *On page 57 it is written that massive stars develop so quickly that it is difficult for astronomers to observe this phase.*

   The time of formation may be very long in years but the locations in the galaxy where such formation is occurring are rare enough that these events are seldom seen.

3. *How are astronomers able to estimate the time at which the first stars formed.*

   Observations of very distant but luminous galaxies called quasars, were made with the Hubble Space telescope. Looking to great distance is the same as looking back to very early time because of the time it takes light to travel the distance to us. Thus we know we are seeing stars that lived in the early universe. The quasar light, which had spent about

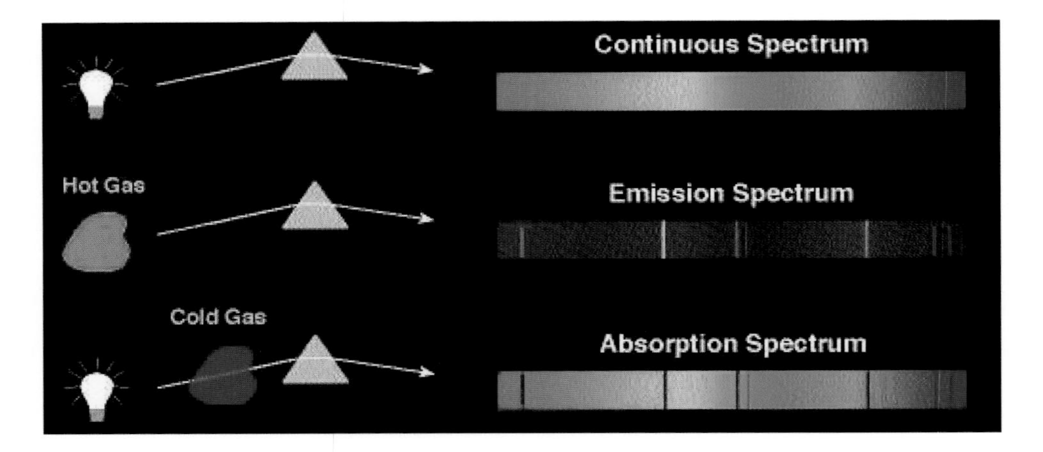

Fig. 3.7 Three types of light spectra are illustrated: (1) A continuous spectrum as from an incandescent light bulb. (2) Line spectrum emitted by a hot gas (the particular lines are *characteristic of all the elements that are present in the gas*). (3) Line spectrum produced when light from a source such as a light bulb passes through a cold gas. The gas absorbs a part of the spectrum producing *absorption* lines that are characteristic of the elements making up the gas.

12.8 billion years crossing space,[b] was split into its different colors, by a laboratory prism which spreads out the colors of light, Figure 3.7. The light from different elements is not a continuous spectrum of colors, but rather appears in lines of various colors in a pattern that is characteristic of each element. Thus the elements that are present in the quasar can be identified, and according to the relative intensity of one element's spectrum as compared to another, their relative abundance can be determined. The light from the first stars contained little iron as compared to recently born stars.

4. How could it be that two-thirds of the atoms in our body were made by the first stars that appeared 200,000 years after the beginning although we are living about 13–14 billion years later? What about all the stars since?

The first stars were very massive and therefore lived a short time, spewing forth their wealth of elements into the cosmos. Later generations of stars were mostly of lower mass and therefore lived longer. Our Sun will live a total of about 12 billion years.

[b]The best measurements to date place the age of the universe at 13.7 billion years with an accuracy of 1 percent (Wilkinson Microwave Anisotropy Probe (WMAP)).

5. *Why have later generations of stars contributed so little to the necessary elements of life?*

Later generations of stars were mostly of low mass. The mass distribution is peaked at low mass and falls rapidly. Low-mass stars live longer than higher mass ones. Our Sun has lived 4.5 billion years and will live a total of about 12 billion years, as already estimated by the great British astronomer and theorist, Sir Arthur Eddington (1882–1944). Low-mass stars like our Sun have made no contribution to the store of vital elements of life like carbon, oxygen, ..., iron. They do not produce elements past helium, and they live so long that they release very little into the cosmos. That is why most of the helium present in the universe is primordial.

3.9. Boxes 11–12

11 Collapse Time of a Gas Cloud

We roughly estimate the collapse time, τ, of a spherical cloud of dust (no pressure) as follows. Here, ρ is the density of the cloud, $\rho = M/((4\pi/3)R^3)$. Equate gravitational potential energy of the outer shell of mass Δm to an average kinetic energy,

$$GM\Delta m/r = (1/2)\Delta m v^2$$

to get

$$\tau = r/v = \sqrt{3/(8\pi G\rho)} \sim 1/\sqrt{G\rho}\,.$$

12 Jeans' Condition for Collapse of a Gas Cloud

Sir James Jeans computed in 1902 the condition under which a gas cloud would collapse. There is a competition between gravity and internal pressure, which rises as the cloud becomes more dense and tends to cause the cloud to oscillate in density. In an otherwise uniform universe of hydrogen and helium, imagine a slight lumpiness here and there. Focus on one of dimension R. As it begins to collapse under its own gravity and its pressure rises, it may bounce, re-expand, collapse, bounce and so on. This oscillation has a period R/u, where u is the velocity of sound in the clump. If the period is greater than the *characteristic* time for the collapse (estimated in the previous box) the clump will collapse; otherwise it will oscillate.

Therefore, gravitational collapse can occur only if the following condition is satisfied,

$$R/u > 1/\sqrt{G\rho}\,.$$

From the value of R given by this condition, we find the volume of the cloud, and from its density we obtain a mass that is referred to as the Jeans mass,

$$M_J = 4\pi\rho u^3/3(G\rho)^{3/2}\,.$$

Chapter 4

Supernovae: Death and Transfiguration of Stars

With all reserve we advance the view that supernovae represent the transition for ordinary stars into neutron stars, which in their final stages consist of closely packed neutrons.

W. Baade and F. Zwicky (1933)

4.1. A Dying Star

Three thousand light-years away, a dying star throws off shells of glowing gas. The Cat's Eye Nebula (Figure 4.1) is one of the most complex planetary nebulae known. Astronomers suspect the bright central object may actually be two stars. The term *planetary nebula*, used to describe this general class of objects, is misleading. Although the objects may seem spherical and planet-like in small telescopes, high resolution images reveal them to be stars that are surrounded by cocoons of gas blown off in the late stages of a star's life. Our Sun also will expand into a red giant engulfing the solar system beyond Mars in blazing gases in about 8 billion years. Such is the relatively tranquil transfiguration of low-mass stars. Heavier stars, a few times our Sun's mass, come to a more violent end — a supernova explosion.

4.2. The Guest Star

"I bow low. I have observed the apparition of a guest star ... Its color was an iridescent yellow ..." So wrote the astonished Court Astronomer of the Sung dynasty in 1054 when he witnessed the sudden appearance of a "guest star." It was clearly visible in daylight and for weeks thereafter. We now know that it had an initial brightness of about 10 billion Suns. The luminosity of the remains of that explosion, the Crab Nebula, now about a thousand years later, is equal to 100,000 Suns. Such sudden flare-ups occur in our own galaxy once every hundred years or so and can be seen even if

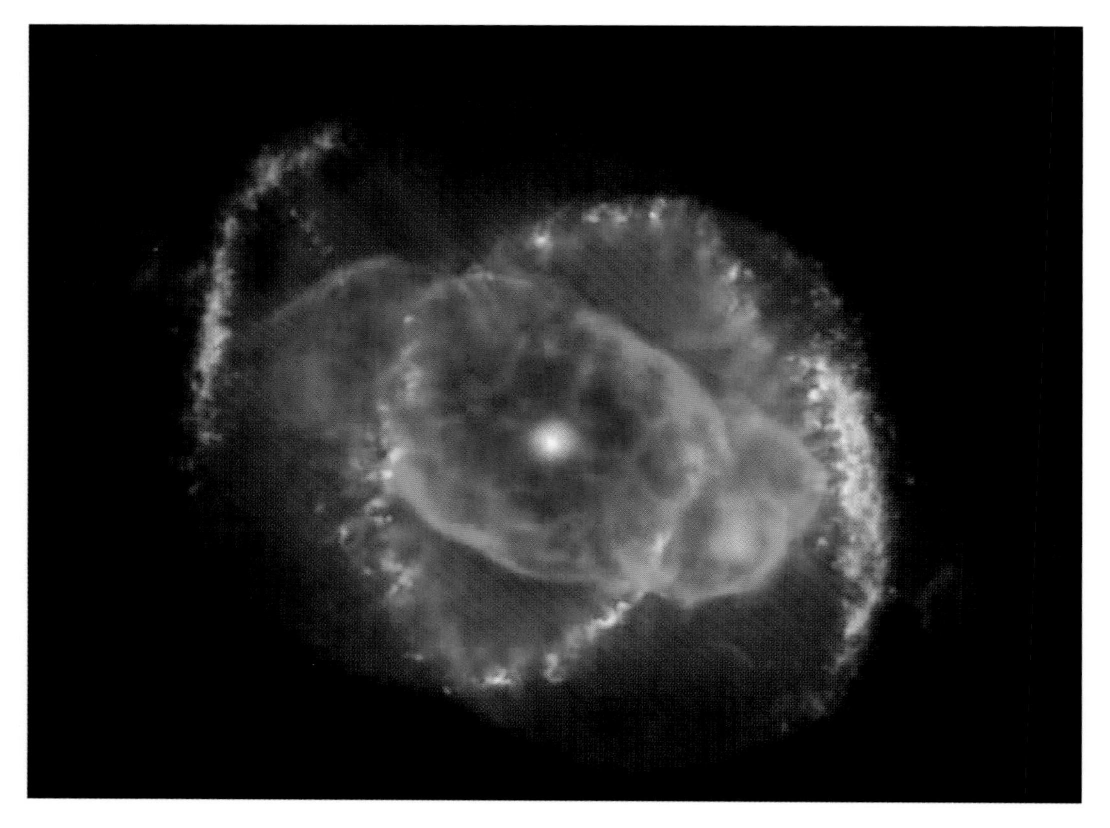

Fig. 4.1 The Cat's Eye Nebula (NGC 6543) is one of the most complex planetary nebulae known. Several stars, instead of the usual one, are in the final stages of their life at its center. They are now ejecting hot gases into space from their flaming surfaces. Surprisingly, the nebula appears to erupt at intervals of 1,500 years, which produces the shells of gases surrounding the dying stars. Hubble images have been taken in 1994, 1997, 2000, and 2002. Credit: J. P. Harrington and K. J. Borkowski (University of Maryland), NASA and HST.

they occur in distant galaxies because of their intrinsic brightness. They came to be called supernova (Figure 4.2).

This event — the explosion of a star in 1054 — was observed by the Court Astronomer of the Sung dynasty and recorded in Chinese history. It seems that it was also witnessed by Anasazi Indians in what is now New Mexico, who commemorated the event with the rock pictograph shown in Figure 1.4.

Nine hundred years later the German-born American astronomer Baade and the Swiss-American theoretician Zwicky, puzzling over these rare but extremely bright outbursts in the distant heavens, advanced the proposition

Fig. 4.2 The arrow at top right points to a stellar explosion — called a supernova — blazing with the light of 200 million Suns. It lies on the outskirts of NGC 2403, a galaxy located 11 million light-years from Earth. Although the supernova is far away, it is the closest stellar explosion discovered in more than a decade. The supernova was part of a compact cluster of stars known as Sandage 96, whose total mass is about 24,000 times the mass of the Sun. Many such clusters — the blue regions — as well as looser associations of massive stars, can be seen in this image. (Supernova are labeled by the year in which they are discovered with a letter following to indicate the sequence of the discovery in that year, as in SN2004a, b, ... z, aa, ab, ...) Credit: NASA, ESA, A. V. Filippenko (University of California, Berkeley), P. Challis (Harvard-Smithsonian Center for Astrophysics), *et al.*

The four forces of Nature

Force	Relative strength	Range (m)
nuclear	10^{38}	10^{-15}
electromagnetic	10^{36}	infinite
weak nuclear (β decay)	10^{32}	10^{-18}
gravity	1	infinite

in 1933 that stars ten times the mass of our Sun or more, end their lives in a cataclysmic supernova explosion that leaves behind the core of the original star, now very much shrunk in size — a neutron star, or sometimes a black hole.

4.3. Core Collapse and Supernovae

It is easy to follow the reasoning of Baade and Zwicky. During its luminous life, a star is supported against gravitational collapse by the pressure created by the heat produced as it combines light elements to form heavier ones: the mass of the produced elements is less than that of the ones from which they are produced. By the equivalence of mass and energy, expressed in the famous law discovered by Einstein — $E = Mc^2$ — the loss in mass is compensated by the production of heat, the pressure of which supports the star. At the end of the life of a star, when it has "burned" all its nuclear fuel in these processes, it is no longer supported against gravity and the star commences to collapse. When the material becomes so compressed that the nucleons comprising the core resist further compression, the core then rebounds, converting some of the gravitational energy into kinetic energy of motion, hurling into space the outer part — and by far most of the star — in what is called a *supernova* explosion. At the same time an enormous shower of neutrinos is produced that actually carries most of the explosive energy. The star's core that is left behind is a very hot neutron star, or, in the case of a very massive star, a black hole. Thus a fiery star with a diameter of a few thousand million kilometers is reduced to one of 15 kilometers, a dead star of "closely packed neutrons" in the words of Baade and Zwicky.

The energy released by the formation of such a neutron star is so great that the explosion can hurl the remains of the original star whose weight is 10 Suns or more into space at a velocity of 10,000 kilometers a second. A thousand years later we see the remains of one such explosion as a diffuse and beautiful object 15 light-years across and still expanding — the Crab

Nebula (Figure 4.3). How the gravitational potential energy gained by the contraction of the star's core was converted into a spectacular explosion belongs to our story of the neutrino.

Fig. 4.3 The Crab Nebula, the remains of a star that exploded in 1054 and was recorded by the Chinese Court Astronomer, who reported to the emperor that the light from the explosion was four times the light from Venus and visible during the day for 23 days. It was probably also recorded by Anasazi Indian artists (in present-day Arizona and New Mexico) on a rock carving. A famous pulsar is located in the nebula, which is a powerful source of energy that is stored in its rotation and slowly released. The present luminosity of the nebula, owing to this powerhouse, is equal to that of 100,000 times that of our Sun. Credit: European Southern Observatory.

4.4. Ghostly Neutrinos

Neutrinos are subatomic particles that interact so weakly with matter that most often a neutrino will pass through the entire Earth without feeling it or being felt. They are produced in great profusion in supernova explosions, the fate that awaits massive stars.

Indeed, most of the gravitational potential energy made available in the collapse of a star to produce a supernova with a neutron star as its residue, would create neutrinos in huge numbers. They would carry with them much more energy than is actually needed to light the heavens and hurl most of the star into space to create the enormous nebula we now see (Figure 4.3). But because of their weak interaction only a small fraction of neutrinos would actually share their energy with the rest of the star and expel it in the supernova explosion. The remainder of the neutrinos, by far the largest number, would escape into space without effect. They would carry with them 10 times more energy than the living star produced in its entire ten million year lifetime. Neutrinos are the messengers from the heart of the newly born neutron star. Just the right number were detected in laboratories on Earth when a star exploded in a neighboring minor galaxy called the Large Magellanic Cloud in 1987 to confirm in broad outline the theory of death and transfiguration of stars in supernovae.

This supernova was the first nearby explosion of a star in three hundred years, Figure 4.4. First to arrive at Earth were the neutrinos. Of these, counters that had been set up at several laboratories in Japan and the United States detected 19. The first light was observed three hours later by Ian Shelton, a University of Toronto research assistant working at the university's Las Campanas station in the mountains of Chile. He began his nightly ritual of taking a three-hour exposure of the Large Magellanic Cloud. When he developed the photographic plate, he immediately noticed a bright star where no star had been seen before. Shelton then walked outside the observatory and looked into the night sky where he saw the vibrant light from a star that exploded 166,000 years ago. This first light arrived three hours after the neutrinos had arrived; it was produced by the decay of excited nuclei in the outer parts of the ejected material of the supernova after the neutrinos had passed at nearly the speed of light. The light display lasted for about 8 months before fading to a quarter of its original brilliance as the last of the long-lived nuclear isotopes produced in the blast decayed.

The Magellanic clouds themselves, in which this explosion took place, are dwarf galaxies seen as large patches of nebulosity that are visible in

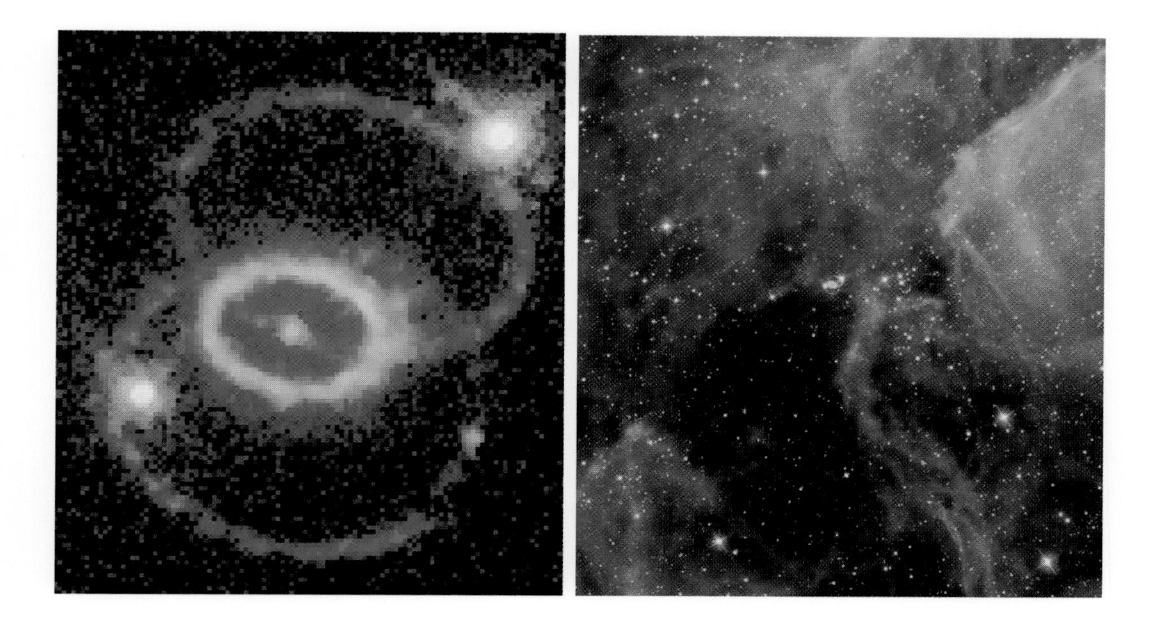

Fig. 4.4 The supernova of 1987A in the nearby Large Magellanic Cloud shown shortly after the explosion, with unexplained rings, and (**at right**) ten years later, surrounded by an enormous nebula of material expelled by the explosion. The rings are still visible a little above center.

the Southern Hemisphere; they appear as though ripped from the Milky Way. These dwarf galaxies are named after the famous Portuguese explorer, Ferdinand Magellan (1470–1521), who observed them on his voyage into the Pacific through the treacherous straits at the tip of South America.

The first known mention of the Large Magellanic Cloud was made by the Persian astronomer Abd-al-Rahman Al Sufi (903–986), working in the court of the Emir Adud ad-Daula in Persia. In his *Book of Fixed Stars* (964 A.D.) he named it the Al Bakr (the White Ox).

4.5. Detecting Far-off Supernova

Supernovae, though the death of stars, also rejuvenate our universe; the shocks they send through space are often the triggers for galaxy formation and star birth. The great tenuous clouds of interstellar gases are occasionally disturbed by some distant supernova explosion and a part of the cloud, compressed by the shock of the explosion, collapses to form a galaxy of stars.

These events typically occur at enormous distance. A close-by explosion would be the end of life on Earth; maybe even of Earth itself. If a close-by explosion occurs, nothing could be done about it. But that is not why astronomers are interested in far-off explosions, which after all do us no harm. They do inform us of some of the wonders of our universe; that, for example, the expansion of the universe is accelerating. They are a part of the great edifice which is our home in the universe. And so with curiosity, we attend its distant reaches.

Neutrinos are the messengers: They are produced in great profusion and travel enormous distances through the universe with no effect on themselves or on whatever they pass through. Therefore it is both an opportunity and challenge for astrophysicist to detect them and to read the information they carry from their distant sources. Astronomers have developed many extraordinary strategies to detect these messengers. They deploy detectors underground, in the Antarctic, and in the oceans. One strategy is to use the Antarctic ice cap as part of a detection instrument (Figure 4.5). It is very deep and its lower parts are very clear and without defects.

The Antarctic Muon And Neutrino Detector Array, referred to as AMANDA, is a telescope for cosmic neutrinos. It is situated at the South Pole. An array of bore holes has been drilled to depths up to 2.4 kilometers in the ice using hot water. Detectors, called photomultipiers, that are very sensitive to light have been attached to long strings and placed in these holes. Though the cosmic neutrino is most likely to pass straight through the Antarctic ice cap, there is the slight possibility that it will react with a nucleus in the ice and produce a muon, a particle similar to the electron, but about 200 times heavier. The muon lifetime is 0.000002 seconds.

Light is slowed in its passage through ice because of its frequent scattering by ice molecules, whereas the muon interacts rarely. Therefore, the muon will travel faster than light in the ice (but of course still slower than the speed of light in vacuum) thereby producing a shock wave of light, called Cerenkov radiation. This light is detected by the photo multipliers, arranged as in Figure 4.5 and when a series of detections have been triggered nearly simultaneously, the path that the neutrinos had followed can be reconstructed with an accuracy of several degrees. In this way the direction of the incoming neutrino, and therefore the direction of the neutrino source, can be pinpointed. The far-off sources of neutrino showers are supernova and galaxies called quasars. (Distant galaxies with bright cores are still sometimes referred to by the name they were given — quasar — before sufficiently powerful detectors revealed the true nature as being a galaxy containing a

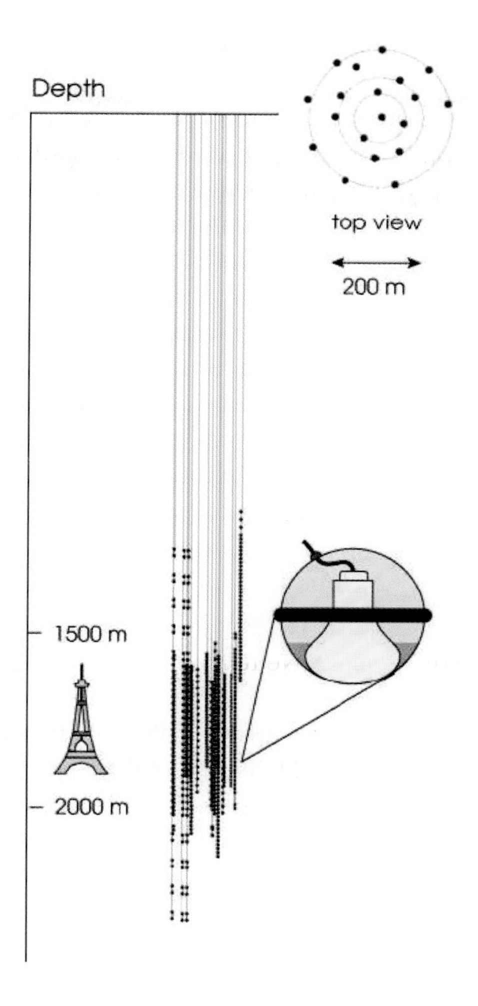

Fig. 4.5 A neutrino telescope deeply buried at 2000 meters in Antarctic ice to shield it from cosmic rays. Neutrino astrophysicists exploit the deep oceans or the 3000-meter-thick Antarctic ice cap. The international group of AMANDA scientists is using this Antarctic ice for detecting neutrinos. At 1500 meters below the surface the ice is exceptionally clear. Light from passing muons can travel hundreds of meters through the ice.

supermassive black hole at the center that is destroying stars as they spiral inward toward their fate.)

The AMANDA collaboration consists of groups from Sweden, the USA, Germany, Belgium, the United Kingdom and Venezuela. The AMANDA project is also supported by the Swedish Polar Research Secretariat and is part of the Swedish polar research program SWEDARP.

4.6. What Sustains a Star So Long Before Its Collapse

The great English astronomer Sir Arthur Eddington, correctly guessed in the 1920s that a star shines because, under the enormous pressure of its own weight, the core is heated to such high temperatures that hydrogen is burned to form helium. He realized that our Sun will shine as we see it now for 12 billion years before consuming all its primordial hydrogen — that the thermal and radiant pressure produced by the fire could resist the attraction of gravity and forestall the ultimate collapse of the star.

Some stars are 10 to 100 times more massive than our Sun. They end their lives in a supernova explosion with a brightness of an entire galaxy having a billion Suns. Their greater weight drives the burning of nuclear fuel more rapidly than stars such as our Sun. Instead of shining for billions of years, such stars burn out in ten million years or so (see table on page 90). Seven generations of (15 Sun mass) stars like these have burned out since the extinction of the dinosaurs 65 million years ago (see page 175). And their greater weight and size causes a whole chain of nuclear burning processes to take place.

What it means for such a star to burn out involves the combined ideas of several scientists seemingly working on unrelated aspects of the lives of stars. So far in this chapter we have described the death of stars: now we describe their afterlife — their transfiguration.

4.7. White Dwarfs and Chandrasekhar

White dwarfs had been discovered in 1910. It was known that they are enormously dense stars and were essentially dead — they produced no new energy but simply radiated what was left from their earlier incarnation. What then could be holding them up against gravity's relentless grasp? This posed a great mystery. But, a young Indian physicist, Subrahmanyan Chandrasekhar (Figure 4.6), who was on his way to work at Cambridge University, realized already on the steamer from Bombay, India (1930), that by the laws of nature a burned out star like a white dwarf could not support much more mass than the Sun. Limits on the mass of a star had never been known before. He derived his limit by combining the Newtonian theory of gravity and the newly formulated quantum theory. The mass limit is now known as the *Chandrasekhar limit*, and is about 1.4 solar masses.

The great British astrophysicist, Sir Arthur Eddington, under whom Chandrasekhar had come to study, dismissed Chandrasekhar's idea as

Fig. 4.6 Subrahmanyan Chandrasekhar as a young man recently arrived at Cambridge University from India by steamer, there to work with Sir Arthur Eddington. While still at sea he solved the mystery that then surrounded white dwarfs — what sustained these dead stars against their own gravity and prevented their collapse.

absurd. From his comments at the time, it is almost certain that Eddington realized that Chandrasekhar's theory of white dwarfs implied continued gravitational collapse to a black hole for those whose mass lay above the Chandrasekhar limit; this seemed unphysical to him. However, most others accepted Chandrasekhar's idea about what sustained white dwarfs. Chandrasekhar went on to a distinguished career, despite the bumpy start, and won many honors including the Nobel prize in 1983 (shared with William Fowler).

Eddington, in his reticence to accept the very concept of a black hole, apparently was unaware of a prescient article — published about 150 years earlier and 200 years before their actual discovery — in the *Philosophical Transactions of the Royal Society* (1783, Vol. LXXIV) by the British cleric, Reverend John Michell (see pages 25 and 92).

4.8. The Pauli Principle

Quantum mechanics plays a direct role in determining the maximum possible mass of white dwarfs (and neutron stars) as discovered by Chandrasekhar. The mass limit for white dwarfs is named after him. He discovered the limit for white dwarfs by applying a quantum mechanical principle discovered by Wolfgang Pauli. According to this principle, not more than one electron can occupy the same state of motion. Chandrasekhar realized that this law had an important consequence for white dwarf stars. In the deep interior of a star the electrons are torn loose from the nuclei of the atoms that are present. The motion of some of the electrons will be very slow, but because of the principle that the young Wolfgang Pauli discovered, the more electrons that are present, the more frantic the motion of some of them. This motion creates a pressure — a resistance to the weight of the matter of the star lying above them. Their pressure sustains the star against gravitational collapse if the star's mass is not too great. Chandrasekhar found the maximum mass for a white dwarf to be 1.4 times the Sun's mass.

The same principle — the Pauli Principle — applies to the dense interior of a *neutron star*, the electrons are forced to combine with protons to form neutrons; hence a star "made of closely packed neutrons" in the words of Baade and Zwicky. Again the Pauli principle applies to the neutrons so that a few of the most energetic ones supply the pressure that resists collapse up to a limiting mass. The limit is similar in value to that of a white dwarf, and is often called the Oppenheimer-Volkoff limit. These authors first estimated its value based on a simple model of a purely neutron star.

However, a neutron star is not made entirely of neutrons. Again, it is the Pauli principle that dictates that. The density of mass, and therefore particles increases toward the center of the star because of the weight of all that lies above them. As the density of neutrons increases, again the Pauli principle comes into play. Each must have its own state of motion, so the more of them there are, the more frantic the motion of some of them: it becomes cheaper, from the point of energy, for a neutron to transform into a proton and electron in a more quiescent state. Together the proton and electron have the same electrical charge as a neutron, thus conserving electric charge, which is one of the laws of nature. Just as with energy, electric charge is conserved. Both may be rearranged, but the sum remains unchanged. Indeed, there are more neutrons in neutron stars than any other

Mass, radius, Schwarzchild radius, and average density of some typical bodies.
(N.s. ≡ typical neutron star; W.d. ≡ typical white dwarf)

Name	M/M_{Sun}	R (km)	r_S (km)	$\rho_{\text{av.}}$ (grams/cm^3)
N.s.	2	10	6	5×10^{14}
W.d.	1	5400	3	3×10^{6}
Sun	1	7×10^{5}	3	1.4
Jupiter	10^{-3}	7×10^{4}	3×10^{-3}	1.3
Earth	3×10^{-6}	6000	9×10^{-6}	5.5

type of nucleon, but many other types of nucleons may be present in the deep interior, and perhaps even quarks [c.f. Glendenning (2000)].

4.9. Neutron Stars

It was a wife and husband team of astronomers Margaret Burbidge and G. R. Burbidge, a famous nuclear physicist W. Fowler and cosmologist F. Hoyle who, in 1957, succeeded in understanding how a star heavier than our Sun would burn ever hotter by consuming a succession of elements, the ashes of one being the fuel for the next. They also realized that the nuclear fuel that keeps a star alive would run out when thermonuclear fusion reactions reached the end of the chain and produced iron. Iron is the most bound of all nuclei so that nuclear burning stops with that element — no energy can be gained by fusing heavier elements from lighter ones. Therefore, a growing core of iron is produced until its mass exceeds the limit calculated by Chandrasekhar. At that point the core would collapse. To what? To a star of "closely packed neutrons" was the suggestion made by Baade and Zwicky. It remained for Colgate and White (1966), and many others (notably A. Burrows) working to this day, to elucidate all the processes by which a fraction of the gravitational binding energy released by the collapsed neutron star, most of it carried away by the weakly interacting neutrinos, could expel the rest of the star in a supernova explosion instead of the whole star falling into a black hole, which no doubt, sometimes happens.

A neutron star, then, is the collapsed core of a very massive star. It is so small in comparison with its original size that there is insufficient room for all the nuclei that were contained in the core. If the size of our Sun were represented by a hundred of these pages, laid end on end, the size of the Earth would be about as long as this page, and the size of a neutron star would be the period at the end of this sentence. Yet more matter fits within a neutron star as exists in the Sun. In such a dense environment, the nuclei of the star are destroyed and in their place there remain the elementary constituents, the neutrons, protons and electrons. Why a star of "closely packed neutrons" as Baade and Zwicky guessed? It was a good guess for the time. The only nuclear particles known were the proton and Chadwick's new discovery, the neutron (1932). From a quantum principle discovered by Pauli (1925), it was easily deduced that most electrons in the star would be captured by protons to form electrically neutral neutrons in the extremely dense environment of a neutron star.

4.10. Discovery: Jocelyn Bell

Neutron stars, though conceived of by Baade and Zwicky in 1933, and studied as theoretical stars by Oppenheimer and Volkoff, and by Tolman in 1939, were not searched for by astronomers, even though the notion of Baade and Zwicky was quite convincing. No one knew what to look for. Such a star, the collapsed core of a once living star, would be quite dead — it could produce no energy of its own, and being very small, would be hard to see by the visible light that it would emit from the store of heat left from its earlier life as part of a large living star. The Reverend John Michell, a visionary of the 1700s had suggested a way of detecting such an invisible star. But he had been long forgotten (see also page 25 and especially 92).

The first discovery of a neutron star was quite by accident. Anthony Hewish of Cambridge had designed a large, and by today's standards, a primitive radio telescope consisting of wires strung out on poles in a five-acre field (see Figure 4.7) for the purpose of detecting signals from very distant objects called quasars.[a] However, soon after it was put into operation, a graduate student, Jocelyn Bell, (Figure 4.9) detected a mysterious signal of a single note, day after day. Some days the signal was stronger, at times weaker, sometimes fading completely — but always returning — and always at the the same interval of time (cf. Figure 4.8). Bell and Hewish (who later won the Nobel Prize), determined that the signal came from outside the solar system. And they suspected, even before announcing their discovery

[a]Today's radio telescopes can be aimed in particularly interesting patches of sky. Hewish and Bell had to wait a day for the Earth to turn to study the same patch.

Fig. 4.7 The 4-acre antennae array of Hewish and Bell, from which the first pulsar signals were discovered in 1967 at Cambridge University. The discovery was serendipitous — pulsars (neutron stars) had not previously been known to exist. Compare this with a modern antenna in Figure 4.12.

to the world in a paper published in the British journal, *Nature*, that they were observing periodic signals from a rotating neutron star.[b]

Coincidentally, Pacini, an Italian working then at Cornell, had published a paper earlier in the same year. He reasoned from one of the conservation laws of physics that neutron stars would have enormous magnetic fields and if rotating, would emit a signal along the direction of the magnet that would be seen once every rotation by an observer who happened to lie in the direction

[b]The Nobel committee has been criticized by some people for awarding the prize in 1974 to Anthony Hewish rather than both to him and Jocelyn Bell. Bell herself has clearly said that she did not agree with these critics, saying that she was a student at the time doing work assigned to her by her supervisor. "It has been suggested that I should have had a part in the Nobel Prize awarded to Tony Hewish for the discovery of pulsars. There are several comments that I would like to make on this: First, demarcation disputes between supervisor and student are always difficult, probably impossible to resolve. Secondly, it is the supervisor who has the final responsibility for the success or failure of the project. ... Thirdly, I believe it would demean Nobel Prizes if they were awarded to research students, except in very exceptional cases, and I do not believe this is one of them. Finally, I am not myself upset about it — after all, I am in good company, am I not!" [full text by Bell can be found on http://www.bigear.org/vol1no1/burnell.htm]

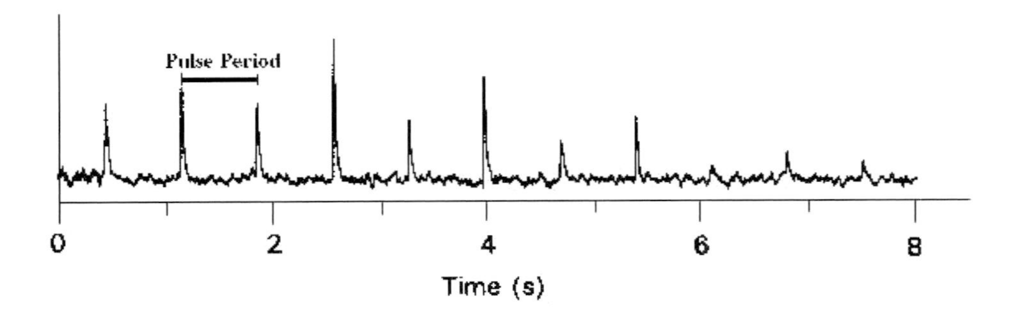

Fig. 4.8 The periodic blips shown are a typical recording of the pulsed radio signal received by large radio antennae from a rotating neutron star (pulsar). The pulses occur at 0.714 seconds in this instance. Image credit: Manchester, R. N. and Taylor, J. H., *Pulsars*, W. H. Freeman, San Francisco, 1977.

Fig. 4.9 Jocelyn Bell whose discoveries of pulsars — rotating neutron stars — while still a graduate student at Cambridge University, electrified astronomers, and opened an entirely new field of astronomy. By timing an orbiting pair of these stars, other astronomers were able to verify Einstein's General Relativity to better the 0.05 percent (page 88).

of the beam. (See Figure 4.10.) The beamed radio signal, rotating with the star just like the beacon of a lighthouse is what Hewish and Bell had detected. They had made the first discovery of a new kind of star — a neutron star.

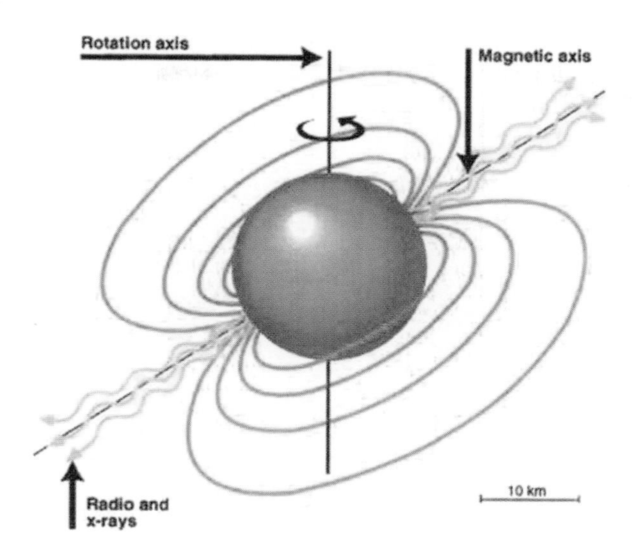

Fig. 4.10 Schematic view of a magnetized neutron star rotating about the vertical axis and with its magnetic axis tilted from the rotation axis. Radio emission beamed in a cone along the magnetic axis is detected as a pulse as the cone sweeps by any observer who lies on the cone. Since most pulsars will not be so oriented that the cone intersects an earthly observer, there are inferred to be many undiscovered pulsars. Credit: the author and LBNL.

The scientific world was electrified. One of the great discoveries in astronomy had been made, confirming earlier ideas about the source of the energy that powered supernovae and opening a new chapter in observational astronomy. It would lead, among other things, to the confirmation of gravity waves as predicted by Einstein's theory of gravity often called General Relativity — the theory, unlike Newton's, that alone can describe gravity where it is very strong as in the region of neutron stars and black holes. As well, General Relativity is the foundation on which the large-scale behavior of the universe from its beginning to its present day accelerating expansion can be described.

Soon other such stars were discovered, including one within the remnant of the star that had exploded in 1054, the Crab supernova (Figure 4.3). Neutron stars that are detected by means of their beamed radiation, which is seen once every rotation, are called *pulsars*. The earthly observer will obtain a pulsed signal made by the beam, if he lies somewhere on the cone that it sweeps out as it rotates (Figure 4.10). Since the discovery of the first neutron star in 1967, some 1600 others have been found through observations made by radio telescopes.

Fig. 4.11 Dick Manchester and Nicolo D'Amico, in the control room of the radio telescope at Parkes in Australia, reviewing evidence of a new pulsar. Manchester leads a large international team from the UK, Australia, Italy and the USA in search for more of the surprises with which these stars have astonished us. Photo by John Sarkissian, Parkes Observatory, CSIRO.

Dick Manchester, (Figure 4.11) who leads a large team of observers, is shown in the control room of a radio telescope at Parkes, Australia. Such telescopes, which are really single, or else arrays of antennae as in Figure 4.12, are tuned to receive the far away and faint emission from the radio waves that a neutron star radiates as it rotates. Radio emission refers to a type of *electromagnetic* radiation. Light is another. They differ in their wavelength as illustrated in Figure 4.13. Our eyes are sensitive to a very small band of wavelengths less than a one-thousandth of a millimeter. Radio waves can be as long as 100 meters and x-rays much smaller than the period at the end of this sentence. Pulsars are usually searched for at wavelengths of a few meters. The difference in wavelength is one reason why our small eyes can see light waves but radio antennae must be large to see radio waves. Large radio receivers for detecting pulsars are associated with universities or research institutes on many continents — Australia, South Africa, Europe

Fig. 4.12 A view of part of the *Very Large Array* in New Mexico. These "dishes" are antennae, tuned to detect the long-wavelength radio waves from pulsars. Credit: NRAO/AUI/NSF.

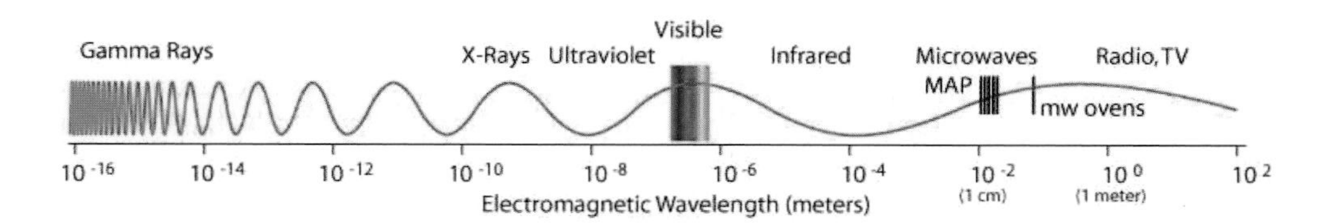

Fig. 4.13 The spectrum of electromagnetic waves, showing the small portion to which our eyes are sensitive compared to the very wide range of other parts of the spectrum from gamma rays to radio waves.

and North America, so that both northern and southern hemispheres of the sky can be viewed.

Some neutron stars rotate only once or twice a second. Even this is very rapid considering that our Sun rotates only once every 25 days. The conservation law of angular momentum assures the spin-up of the collapsing core of a slowly rotating star, just as an ice skater can spin up by drawing in his outstretched arms. The young Crab pulsar, the very neutron star whose formation created the supernova of 1054, spins 33 times a second. A star the mass of the Sun spinning at this rate has an enormous store of energy. It is known from observations made on the nebula that the pulsar within is converting its rotational energy into radiation at a power equal to that of 100,000 Suns. This energy lights the Crab Nebula, 15 light-years in diameter, making it luminous in radio, optical and x-rays, and accelerates wisps of gas in it to half the light speed. The nebula, consisting of the 15 solar masses of debris from the star that exploded in 1054 will continue to expand becoming more diffuse and dimmer. Perhaps in 50,000 years it will disappear while the pulsar, only little reduced in spin, will join the other isolated pulsars that are known to exist through the discoveries of the last forty years. It will continue its outpouring of energy for another 10 billion years, but at a decreasing rate as its rotation slows [Glendenning (2000)]. Finally it will disappear from the radio sky and will join a growing number of silent unseen neutron stars.

The first very rapidly rotating pulsar was discovered in 1982 by Donald Backer. It rotates 642 times per second. An even faster one has very recently been discovered that is rotating 716 times a second. Considering that an object with the mass of our Sun, or somewhat more, with a radius of only 15 kilometers is spinning hundreds of times per second as compared to the Earth's rotation of once in 24 hours, or the Sun which takes 25 days to rotate once, one realizes the enormity of the energy stored in the pulsar's rotation. In 10 million years — long after any civilization on Earth as we know or can conceive it, has passed — this pulsar's signal will still be blinking toward our planet — perhaps then lifeless because of the destruction of the ozone layer by industrial pollution or other possible catastrophes.

Some neutron stars move with enormous speed through space, their trail marked by a disturbance (ionization) of the interstellar gas as shown for the so-called "mouse" in Figure 4.14: it is moving through space at 1.3 million miles per hour.

Today, at a few laboratories in both hemispheres, giant radio telescopes (Figure 4.12) are trained at the sky in a hunt for these strange neutron stars. It is believed that as many as 30,000 of them inhabit our Galaxy. Neutron stars, with one or two exceptions, make themselves "visible" to astronomers by a focused radio beam, which, because all stars rotate, appears to the earthly astronomer like a pulsed signal — a lighthouse (Figure 4.10). Hence the name pulsar. Since the first discovery in 1967, there have been many

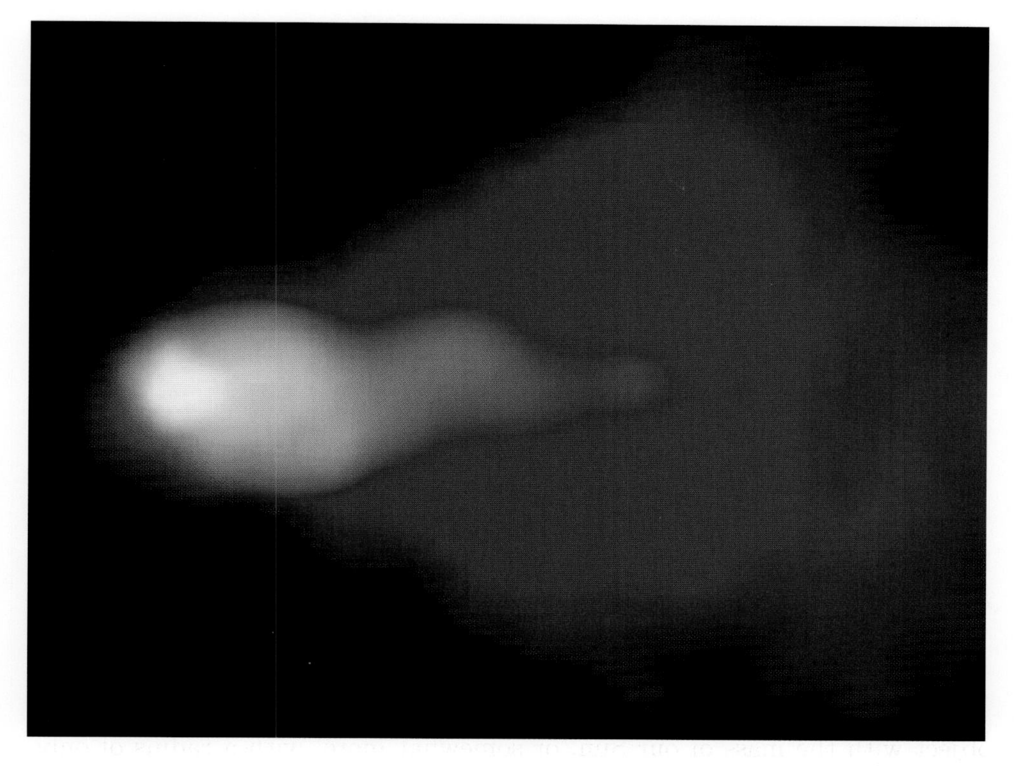

Fig. 4.14 A nebula in flight called the Mouse, from its appearance in radio images that show a compact snout, a bulbous body, and a remarkable long, narrow, tail that extends for about 55 light-years. The X-ray cloud consists of high-energy particles swept back by the pulsar's interaction with the interstellar gas. An intense X-ray source near the front of the cloud marks the location of the pulsar, estimated to be moving through space at about 1.3 million miles per hour. Credit: NASA/CXC/SAO/B. Gaensler *et al.* Radio: NSF/NRAO/VLA.

surprises; the first pulsar with a Sun's mass spinning an incredible 642 times a second discovered by D. Backer from UC Berkeley; the first extra-solar planetary system — three planets around a neutron star — discovered by Alex Wolszczan from Penn State and Dale Frail from the National Radio Astronomy Center, Socorro, New Mexico. Wolszczan and Frail used the world's largest radio telescope in Puerto Rico (see Figure on page 208) to detect signals coming from a distant tiny star in the constellation Virgo, 7,000 trillion miles from Earth.

Many of these discoveries — besides revealing new wonders of our universe — have important consequences, none more so than the discovery by Russell Hulse and Joe Taylor in 1973 of a pair of neutron stars rapidly orbiting each other. What is most important is that one of them is a pulsar whose beam intersects the Earth. By taking measurements on their orbital motion, as revealed by the pulses, over a period of 25 years, Einstein's General Relativity and the reality of gravitational radiation were confirmed to the 1 percent level. For this work they shared the 1993 Nobel Prize in Physics. R. N. Manchester, A. Lyne, and N. d'Amico at the radio telescope in New South Wales, Australia, have recently discovered a pair of orbiting *pulsars*. With beams from both neutron stars, they have already accumulated enough data in two years to confirm General Relativity predictions to 0.05 percent level.

4.11. Birth, Death, and Transfiguration

How does nature make these strange objects? That is the story of the birth, death, and transfiguration of stars. The matter in the very early universe — before there were stars — was in the form of a diffuse gas of very light elements, about four parts of hydrogen and one of helium. The very lightest elements were formed in the first few minutes in the life of the universe. The heavier elements so important for life, like carbon, oxygen, and iron, had not yet been created, and did not make their first appearance until the earliest giant stars were born and died, 200,000 years later. Under the attraction of gravity, vast clouds of diffuse gas were pulled together to form galaxies filled with stars. Stars are still being formed in great clouds of gas (such as the beautiful column shown in Figure 4.15) that occupy some of the space between the stars in our own galaxy, the Milky Way.

Stars like our Sun are great furnaces. Our Sun provides the warmth and energy by which we live. And like all furnaces, stars consume their fuel and burn out. The process known as thermonuclear fusion, fires these furnaces. *Thermo* refers to heat, *nuclear* to nuclei, and *fusion* to the fusing of light elements into heavier. Three helium nuclei are fused in a star to form carbon; its mass is less than the three helium nuclei. According to Einstein's famous law $E = Mc^2$, the loss in *mass* appears in another form of energy — *heat*. Some of this heat energy is radiated into the universe, but the pressure exerted by the rest of it resists the force of gravity from causing the immediate collapse of the star. A whole sequence of such burnings takes place involving the production of ever heavier nuclei by the fusion of lighter ones — four hydrogen nuclei to produce helium, three helium to produce carbon and so on, until iron and nickel are made. But each successive reaction produces less energy and therefore less pressure than the preceding one. Consequently the star slowly contracts under gravity's grip. The process ends with the production of iron, because for elements heavier than iron,

Fig. 4.15 Parts of this great cloud of gas — called the Eagle Nebula, many light-years across — are being pulled together by gravity to form stars. Some of the star cradles are seen as bright nodules near the top of the column. Credit: Jeff Hester and Paul Scowen (Arizona State University), and NASA.

fusion does not produce energy; an *infusion* of energy is required to build heavier elements.

In low-mass stars, because gravity is weak, these processes proceed very slowly. Our Sun will live for 12 billion years. But more massive stars — like the one that produced the only visible supernova in our lifetime — live for only about 20 million years.

The iron that is produced as a star burns, sinks to the center and a limiting mass is finally reached — the Chandrasekhar limit — about 1.4 times our

Evolutionary Times of Stars

M/M_{Sun}	Life (Yrs)	Eventual Fate
30 or more	5×10^6	Black Hole
15	1×10^7	Neutron Star
10	2×10^7	Neutron Star
5	7×10^7	White Dwarf
1	1×10^{10}	White Dwarf
0.1	3×10^{12}	White Dwarf

Sun's mass. At this point the inert iron core of a thousand kilometer radius can no longer resist its own gravity — it collapses in a fraction of a second to a 10 kilometer very hot proto-neutron star leaving behind the rest of the star which, with no support from below, then begins to collapse. But a burst of neutrinos is emitted by the hot core of the collapsed star in its final death throes. They carry ten times as much energy as the luminous star released in its entire lifetime of tens of millions of years. Even a small part of this energy is far more than enough to blow the rest of the star apart, and cast it off into space at 10,000 kilometers per second. The explosion is called a nova or a supernova. Supernovae and Novae — the death of stars and birth of neutron stars or white dwarfs — can be seen even in very distant galaxies at the very limits to which we can presently see with powerful telescopes. At the peak of the explosion, the brightness equals tens of millions of Suns, rivaling the brightness of their host galaxy. Even now the Crab Nebula, a thousand-year-old supernova remnant, Figure 4.3, is as bright as 100,000 Suns.

Only a few of the lightest elements were produced during the primordial nucleosynthesis in the beginning when the universe was intensely hot and dense, none heavier than about lithium. Other elements, up to iron, are synthesized by thermonuclear reactions during the 10^7 year evolution of massive stars, the heavier of these in the last few days of the life of the presupernova star. The heaviest elements are synthesized after the supernovae explosion just outside newly formed neutron star in the material of the rest, and by far the most of the dying star, which suffers an intense neutrino bombardment as the hot neutron star releases its binding energy.

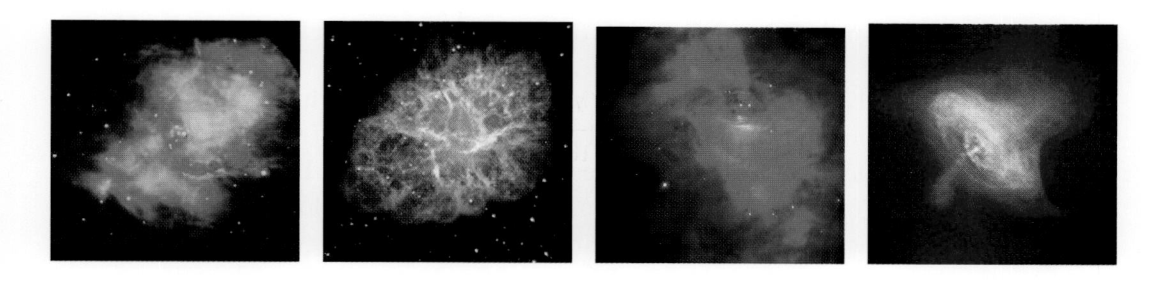

Fig. 4.16 Four different views of the Crab supernova remnant. The first and last are taken at infrared and x-ray frequencies, the center two at optical, one with a ground-based telescope, the other with the Hubble Space Telescope showing the central region of the Crab Nebula. The neutron star is at the center of a sort of whirlpool diagonally to the right and up. The fourth view, at x-ray frequency, is also a view of the central region that reveals one end of a jet expelled by the neutron star rotating 30 times a second. The other end is faintly visible and apparently deflected, possibly by a shock wave from another explosion far away. Credit: NASA.

These elements diffuse into the cosmos there to wander for eons, mixing with the gases of myriad other exploding stars and with the primeval clouds of hydrogen and helium to begin once again the cycle of star birth, transfiguration, and death. Each generation of stars contains a richer mixture of the heavier elements. After many generations enough dust and heavy elements had accumulated in the host galaxy to form the first planets in orbit about Suns like ours. On at least one of these, from the stuff of stars, life arose.[c] About one hundred generations of stars have lived and died since the formation of our galaxy. So here we are to marvel at our own existence and the wonders of the universe.

But — occasionally, the neutrino burst does not cast away all of the outer part of the dying star. In this situation it falls back onto the neutron star, adding to its mass, sometimes to the point of exceeding the maximum that can be supported by the internal pressure against gravity. A black hole is then formed from which nothing can ever return. Very massive stars — 50 times the Sun's mass — are believed always to create black holes. In these events, a great quantity of the elements that were manufactured during the lifetime of the presupernova are lost forever to the cosmos.

[c]More than 130 other solar systems with planets are known to exist in our own galaxy.

4.12. Reverend John Michell and the Idea of Black Holes

John Michell (1724–1793) was born in Nottinghamshire three years before Sir Isaac Newton's death. He studied at Queens' College of Cambridge University where he received an MA degree in 1752, and a BD (Bachelor of Divinity) in 1761. In the same year he was elected Fellow of the Royal Society, and briefly held the Woodwardian Chair of Geology at Cambridge. Shortly thereafter he was appointed rector of St. Michael's Church of Thornhill, near Leeds, Yorkshire, England, a post he held for the rest of his life. Reverend John Michell died there in 1793 at age 68.

As a geologist, Michell constructed a torsion balance for measuring gravitational forces. In 1760, he constructed a theory of earthquakes as wave motions in the interior of the Earth, and suspected a connection between earthquakes and volcanism. In fact, the 2005 tsunami in Southeast Asia was caused by volcanic eruptions off the coast of Indonesia, thus confirming Michell's 200-year-old speculations.

In 1767 he published an investigation on double stars and clusters, and calculated the probability of finding chance alignments of stars (asterisms). In particular, he investigated the Pleiades cluster, and calculated a probability of 1/496,000 to find such a group as a chance alignment anywhere in the sky. Also, he found that far too many pairs and close groups of stars were visible in the sky to assume that all these were chance alignments. and concluded that many of them should be physical pairs or groups, held together by ... their mutual gravitation. Indeed, astronomers now know that more than half the stars reside in groups.

As an amateur astronomer, John Michell was also an active telescope builder. His main instrument was a self-made 10-foot [focal length] reflector of 30-inch aperture. This instrument was purchased by William Herschel after Michell's death, and served as a model for a similar instrument. The work of Michell had some impact on William Herschel, and in particular, stimulated his work on double stars.

Most amazing to us today, The Reverend Michell had already conceived of *black holes* and how they might be discovered in 1783; "... if there should really exist in nature any bodies, whose ... light could not arrive at us [because of its strong gravitational field] ... yet, if any other luminous bodies should happen to revolve about them we might still perhaps from the motion of these revolving bodies infer the existence of the central ones with some degree of probability"(in a letter to John Cavendish in 1783 and published in the *Philosophical Transactions of the Royal Society* in the following year).

In fact, this is exactly how modern astronomers have detected the existence of many black holes of a few solar masses.[d] In fact he discovered the relationship between radius and mass that defines the black hole horizon (see Question 2 below).

4.13. Questions

1. *Neutrinos have very small mass. Therefore they can travel near, but not at, the speed of light. Why therefore did they arrive at Earth before light from the supernova in the Magellanic cloud?*

Because the photons of light were scattered, captured, reemitted, captured ... many times by the atoms of the debris expelled in the explosion.

2. *Calculate the Schwarzschild radius of a black hole in the manner in which the Reverend John Michell reasoned, 200 years ago.*

Equate kinetic energy of escape to gravitational energy, setting (incorrectly) the escape velocity of mass to light velocity, c: $(1/2)mc^2 = GMm/R$ to get $R = 2GM/c^2$, which is the Schwarzschild horizon radius R of a black hole of mass M, from whose interior not even light can escape, as derived in General Relativity. That a classical calculation should give the correct answer must be considered as fortuitous.

3. *Why is a number such as 5 solar masses chosen by astronomers as the minimum mass for a black hole "candidate"?*

The theoretical lower limit for a black hole "candidate" is chosen so that it lies comfortably above the maximum possible mass of a neutron star. This is not precisely known. However based on minimal assumptions about the relation between pressure and density of neutron star matter (called its equation of state) the maximum mass is about 3 solar masses. Calculations of the maximum mass of neutron stars that are based on more realistic theories of dense matter are around 1.5–2 solar masses.

[d]It would not be possible to distinguish between unseen neutron stars or white dwarfs and low-mass black holes by only their gravitational effect on a neighboring luminous star. Therefore in identifying "candidates" for black holes, the usual convention is to require that their mass be at least 5 solar masses which is above the maximum possible mass ($\approx 3M_{\odot}$) of a neutron star or white dwarf.

Chapter 5

Nebulae

Nothing exists except atoms and empty space; everything else is opinion.

Democritus (470–380 B.C.)

5.1. The Milky Way and Nebula

The Milky Way, that band of light we can see across the night sky is our own galaxy, a vast collection of stars. It has the form of a thin disk and a central bulge. Our Sun is in the disk and about two-thirds of the way from the center. When we view the Milky Way from our home on Earth, we are looking back toward the center where most of the stars are. There is also a great deal of gas and dust that has been ejected from the hot surfaces of stars. The dust obscures the view to the other side of the galaxy. The dust is not uniform: We can often discern picturesque forms, which in our imagination we translate into tableaux such as the Horse Head Nebula (Figure 5.1).

Nebulae are the denser regions of molecular clouds in our own galaxy, clouds containing mostly hydrogen and helium with some molecules that have been cast off by stars. Some nebulae glow from the light emitted by nearby stars, often newly born stars; they are called *emission nebulae* like the *Rosette* seen in Figure 5.2, and the *Pelican* in Figure 5.3. Others, like the *Horse Head*, and the Witch Head (see preface), are visible against the background of illuminated dust and are called *reflection nebula*, while others like the *Fox* seen in Figure 5.4 have both features. Other distant nebulous objects observed more than a thousand years ago by the Persian court astronomer Al Sufi turned out to be far off galaxies, the "Island Universes" observed by William Herschel through his telescope in the late 1700s. No doubt many, but not all, of these distant galaxies also contain nebulae, but they would be too faint to detect. Those galaxies that have experienced a near collision with another are likely to have been swept clean of gases, and so contain no nebulae.

Fig. 5.1 The Horse Head Nebula, a pillar of gas in a great cloud. The red glow originates from hydrogen gas predominantly behind the nebula, ionized by the nearby bright star Sigma Orion. Bright spots in the Horse Head Nebula are young stars in the process of forming. Credit: Mauna Kea, Hawaii with the Canada–France–Hawaii Telescope.

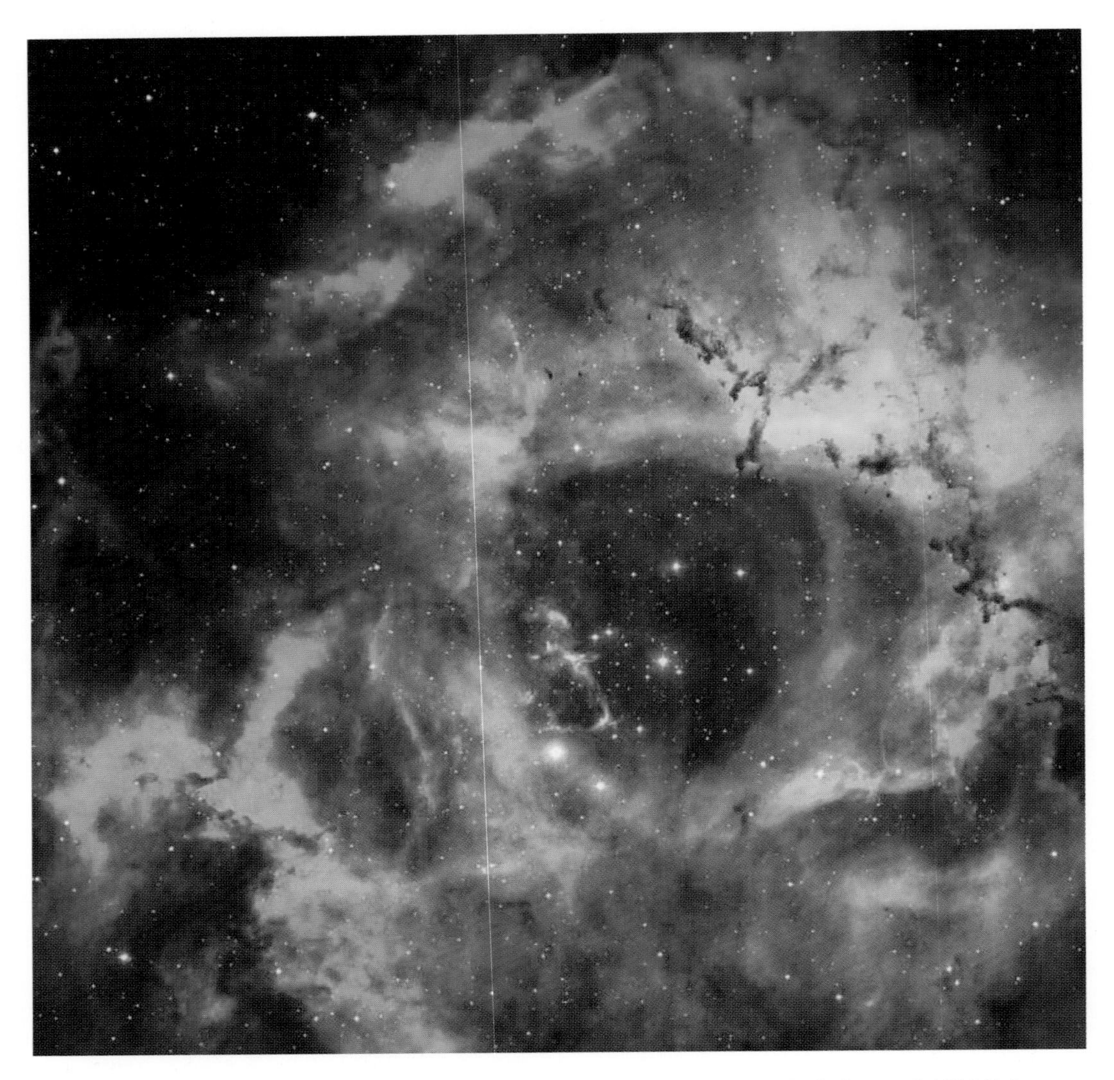

Fig. 5.2 The Rosette Nebula (NGC 2237) is a large emission nebula located 3000 light-years away. The great abundance of hydrogen gas gives NGC 2237 its red color. The wind from the open cluster of stars known as NGC 2244 has cleared a hole in the nebula's center. This photograph was taken in the light emitted by three elements of the gas ionized by the energetic central stars. Here green light originating from oxygen and blue light originating from sulfur supplements the red from hydrogen. Filaments of dark dust lace run through the nebula's gases. The origin of recently observed fast-moving molecular knots in the Rosette Nebula remains a mystery. Credit: T. A. Rector, B. Wolpa, M. Hanna (AURA/NOAO/NSF).

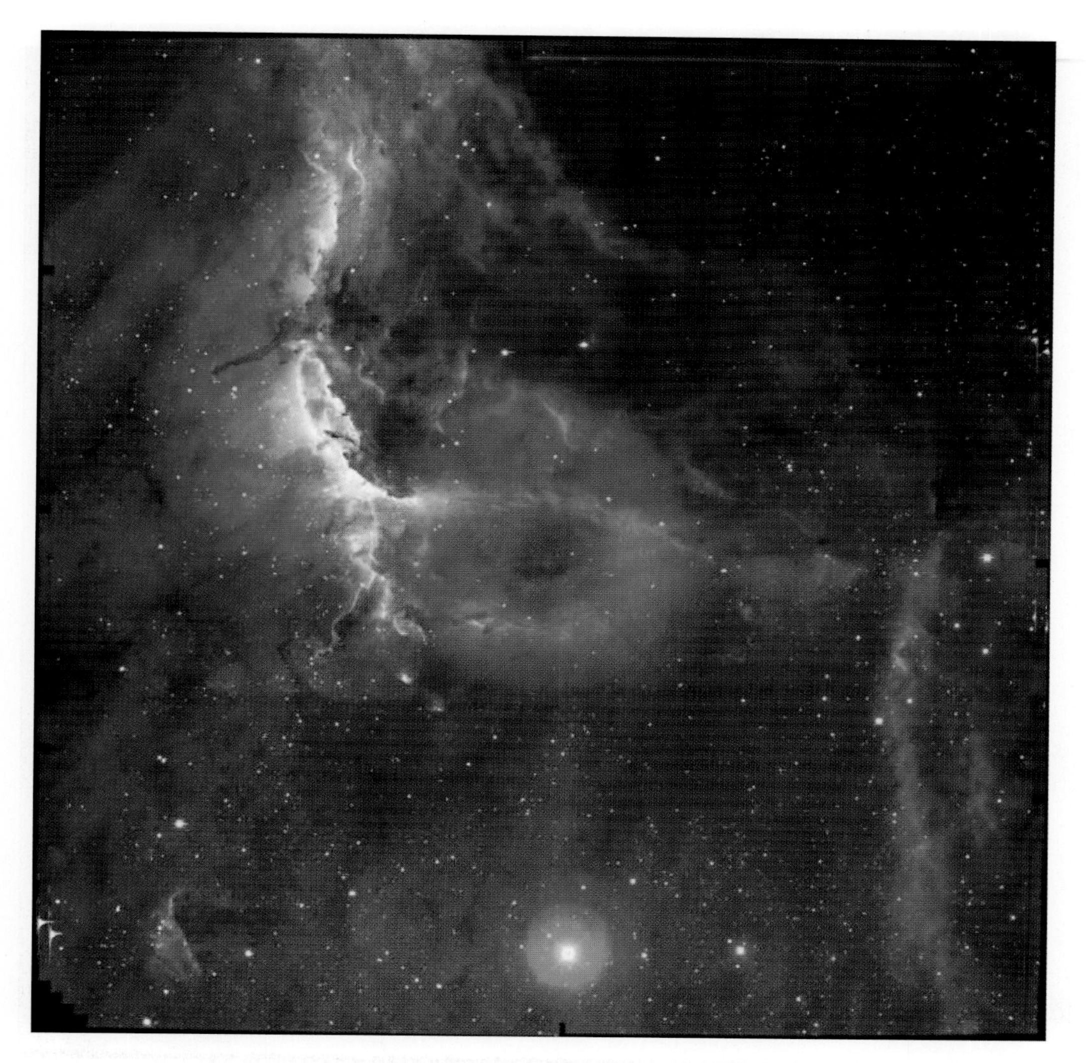

Fig. 5.3 The Pelican Nebula, in the red light of recombining hydrogen atoms and singly ionized sulfur, reveals a population of shock waves that trace outflows from stars that are in the process of formation; they are embedded within the molecular clouds that rim the nebula. The visibility of these outflows is enhanced by the ultraviolet radiation of several massive stars located off the image. This radiation is eroding the surrounding molecular cloud. In regions shadowed by dense clumps of gas and dust, parts of the dense and cold molecular cloud survive to produce the long pillars of dusty material. A faint jet (top left quadrant and to the left of the long vertical bright line) squirts out of the tip of one of the pillars, apparently indicating the presence of an unseen protostar. Credit: John Bally (U. Colorado), Bo Reipurth (U. Hawaii)/NOAO/AURA/NSFs.

Fig. 5.4 The Fox Nebula has an appearance reminiscent of ancient Chinese paintings. The strange shape originates from fine interstellar dust reacting in complex ways with the energetic light and hot gas being expelled by young stars. The blue glow results from reflection, where neighboring dust reflects light from a bright star. The more diffuse red glow results from emission, where starlight ionizes hydrogen gas. Pink areas are lit by a combination of the two processes. This nebula is near NGC 2264, located about 2500 light-years away toward the constellation of Monoceros. Photo ©David Malin, Anglo-Australian Observatory.

5.2. Sir John Herschel and the Carina Nebula

John Herschel was the son of the great Sir William Herschel, the astronomer who first discovered that the distant nebulous objects, which he could see with his telescope, were galaxies lying far beyond the Milky Way. He called them *Island Universes.*

The father, Sir William, who immigrated to England from northern Germany at the age of 21 to escape the war with France, became not only a leading astronomer and eventually Astronomer Royal to King George III, but he also had a great talent for music. Indeed, during his early days in England he was a music teacher and organist at the Octagon Chapel in Bath. His sister too, Caroline Herschel, was extremely musical, and they both used their musical talents to help augment their income after first arriving from Northern Germany. John was brought up in his father's home, Observatory House, with its 40-foot telescope, where music, science and religion were harmoniously mingled. His aunt, Caroline Herschel (Figure 1.3), Sir William's sister, was an astronomer in her own right, having at first assisted her brother, William, in his nightly search of the heavens. She became the first woman in England to be recognized as a scientist with the award of a salary from King George III and eventually was recognized throughout Europe as a leading astronomer for which she reaped many honors including the gold medal of science by the King of Prussia.

The son, John,[a] was a multifaceted genius who reaped many honors in his day. As an undergraduate at Cambridge, John Herschel and two friends founded the Analytical Society with the aim of introducing methods of mathematical analysis used by the great French mathematicians, d'Alembert, Leibniz, and Lagrange into English universities. In 1813 he was elected as a fellow of the Royal Society of London, and published papers and a two-volume book on mathematics in 1820.

Herschel's remarkable all round abilities and pursuit of many topics — law, photographic chemistry, and mathematics — apparently diluted his contribution to any one of them so that he failed to make an advance of the depth that he was clearly capable of in any of the subjects that he studied.

While spending a holiday with his father in the summer of 1816, he decided to turn to astronomy, almost certainly influenced by the fact that at 78 years of age his father's health was failing and there was nobody else to continue his work. Although John Herschel took up astronomy from this

[a]These notes on John Herschel are based on a biography by J. J. O'Connor and E. F. Robertson; http://www-history.mcs.st-andrews.ac.uk/Mathematicians/Herschel.html.

time, he also studied other topics. Even before his first astronomy paper was published, Herschel published details of his chemical and photography experiments in 1819, which years later, proved of fundamental importance in the development of photography. He was very much involved with the founding of the Astronomical Society in 1820 and he was elected vice-president at the second meeting.

Herschel's involvement with the Royal Society had important influences on his career. He had been proposed for President of the Royal Society. But in an embarrassing episode for Herschel, he failed narrowly to be elected. However, he was elected as President of the Astronomical Society in 1827 and he was knighted in 1831.

Nevertheless, the Royal Society episode may have been the main reason why Herschel decided to make a long visit to the Cape of Good Hope in South Africa. There, he made important observations, including an account of Halley's comet on its 1835 return. Herschel recognized that the comet was being subjected to major forces other than gravity and he was able to calculate that the force was one repelling it from the Sun. This could in some sense be said to constitute the discovery of the solar wind, which is indeed the reason for the repulsive force discovered by Herschel. He also made the important discovery that gas was evaporating from the comet. His observation of the *Carina Nebula* (sometimes referred to as the keyhole, Figure 5.5.) was also made in South Africa.

John Herschel's observations were made before photography was used together with telescopes to obtain images of heavenly objects, and so he made a beautiful and careful painting of the Carina Nebula. "Carina is home to Canopus (alpha Carinae), the second brightest star in the heavens. Its name is said to come from the pilot of the fleet of ships of King Menelaus. It was this king of Sparta who rallied the men of Sparta to fight for Helen of Troy; his prize was Helen herself, who became his queen. Of Canopus, the pilot, it is said that he died in Egypt after the fall of Troy."[b]

In January 1839 Herschel heard of the Frenchman, Daguerre's work on photography from a casual remark in a letter written to his wife. Without knowing any details, Herschel was able to take photographs himself within a few days. "I think you'll find that fixed." This was the beginning of the hyposulphite method of fixing.

Indeed Herschel was able to achieve this remarkably rapid breakthrough due to the work that he had conducted and published on chemical processes related to photography. Many people have wondered why Herschel himself

[b]Richard Dibon-Smith (www.dibonsmith.com).

Fig. 5.5 The Carina Nebula (NGC3372), with an overall diameter of more than 200 light-years, is a prominent feature of the Southern-Hemisphere portion of the Milky Way. The speeds of the interstellar clouds range from over 500,000 mph toward the observer to nearly 300,000 mph away from the observer. The nebula is adjacent to the famous explosive variable star Eta Carinae. The Carina Nebula also contains several other stars that are among the hottest and most massive known, each about 10 times as hot, and 100 times as massive, as our Sun. This nebula was described and drawn by Sir John Herschel, observing from South Africa just before the middle of the nineteenth century. Credit: K. Weis and W. J. Duschl (ITA, University of Heidelberg).

never made the steps which would have led to his being recognized as the inventor of photography. There was a period of around 20 years from the time of his 1819 work when he might have made the breakthrough but again it was probably due to his wide ranging talents that he failed to do so. Not only did his talents take him into a wide range of other activities but his great skill as an artist meant that he had less need to invent photography than most others.

At the age of 58 Herschel made a rather strange decision as to the future direction of his career. He had turned down entering parliament as a Cambridge University member and he had also turned down a proposal that he become president of the Royal Society. He accepted instead the post of Master-of-the-Mint. In this he followed in the footsteps of Sir Isaac Newton 150 years earlier. And like Newton, he no longer pursued his scientific research. Although he made no major breakthroughs for which he is remembered today, he was considered by many of his contemporaries as the leading scientist of his own day.

5.3. Planetary Nebulae

Planetary Nebulae actually have nothing to do with planets. Before astronomers knew what they were, they attached the adjective "planetary" because of the central dot that suggested a planet. Nebulae are actually much too far away for a planet to be observed. Instead, the central object is a dying star that is throwing off its outer layers of hot gases in its final death throes. Light from the star illuminates them, forming what is called a *planetary nebula*.

A planetary nebula forms when a star can no longer support itself by fusion reactions in its center. Gravity squeezes the inner part of the star by the weight of the overlaying material, raising the temperature thereby. Radiation from the high temperature central core drives the outer part of the star away by what is called a *stellar wind*, lasting a few thousand years. Afterward, the remaining core thus laid bare, radiates heat to the now distant gases, causing them to glow.

Smaller stars like our Sun, and heavier ones up to five times our Sun's mass, end their lives in this way. Toward the end, the dying star begins a series of puffing expansions into a red giant, loses heat, and therefore thermal pressure, and then re-contracts under the force of gravity. During each such cycle, some of the outermost shells of gas escape into space. Eventually, after sufficient loss of gas, gravity fails to halt the expansion and the outer parts become detached, to form a planetary nebula. The remaining stellar-core remnant, uncovered in this way, heats the now distant gases over time and causes them to glow such as the beautiful *Cat's Eye* (Figure 4.1), the *Dumbbell* (Figure 5.6), and the *Eskimo* (Figure 5.7).

The escaping gases from the nebula contain the important elements, carbon and oxygen, which disperse and wander through the universe, some few atoms of which may be contained in our own bodies and in the air we breathe. A hot X-ray emitting white dwarf, lying at the center of the nebula, is all that remains of the once shining star. It will slowly cool over centuries.

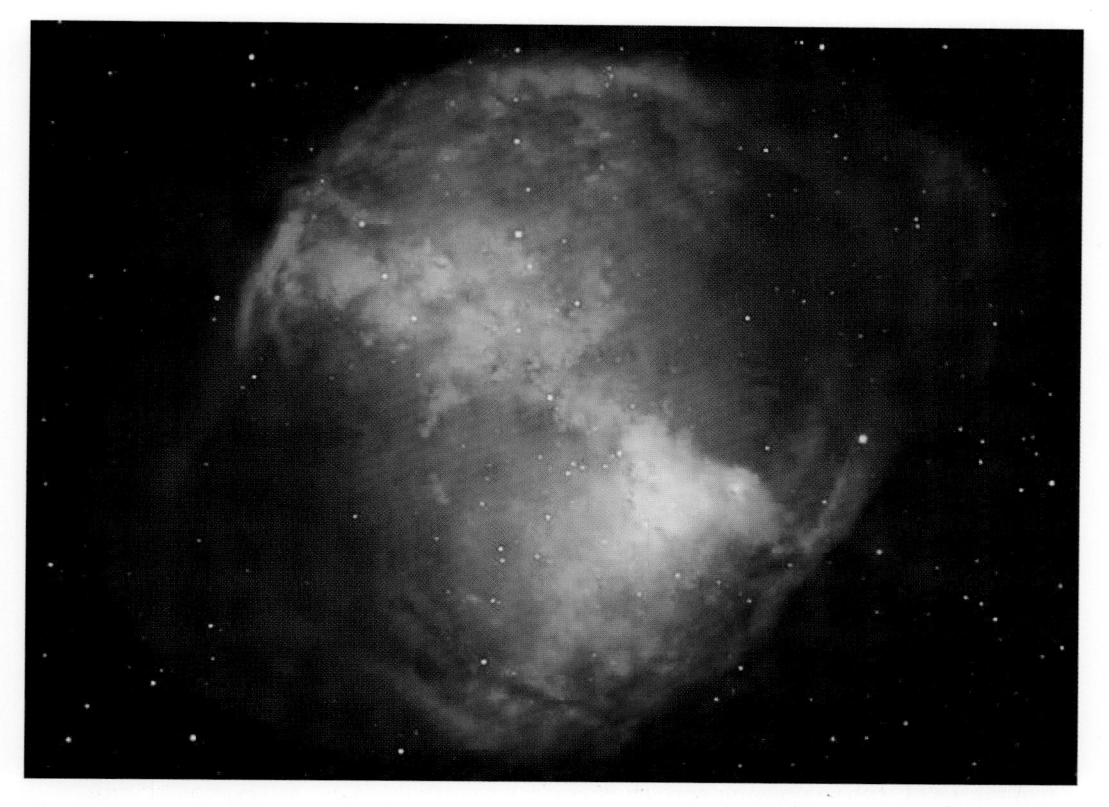

Fig. 5.6 The Dumbbell, M27, is one of the brightest planetary nebulae in the sky, and can be seen in the constellation Vulpecula with binoculars. It takes light about 1000 years to reach us from M27, shown above, digitally sharpened, in three isolated colors emitted by hydrogen and oxygen. Credit: NASA, ESA, HEIC, and the Hubble Heritage Team (STScI/AURA).

5.4. The Cerro Tololo Observatory

Some astronomical images are taken from satellites while others are taken from ground-based telescopes, often perched high above industrial and city-light pollution, such as the Cerro Tololo Observatory in the mountains of Chile, as seen in Figure 5.8. The Chilean government passed a strong *lighting law* (*DS686*) in December 1998, ordering that all lighting fixtures in regions of Chile (which includes all the present and potential astronomical sites) be changed for fixtures which prevent light from shining directly into the sky. The changeover process is underway; 30 percent of all fixtures have already been changed and will be completed by October 2005. The only powerful ground-based telescope in the U.S. that can penetrate to great distance in

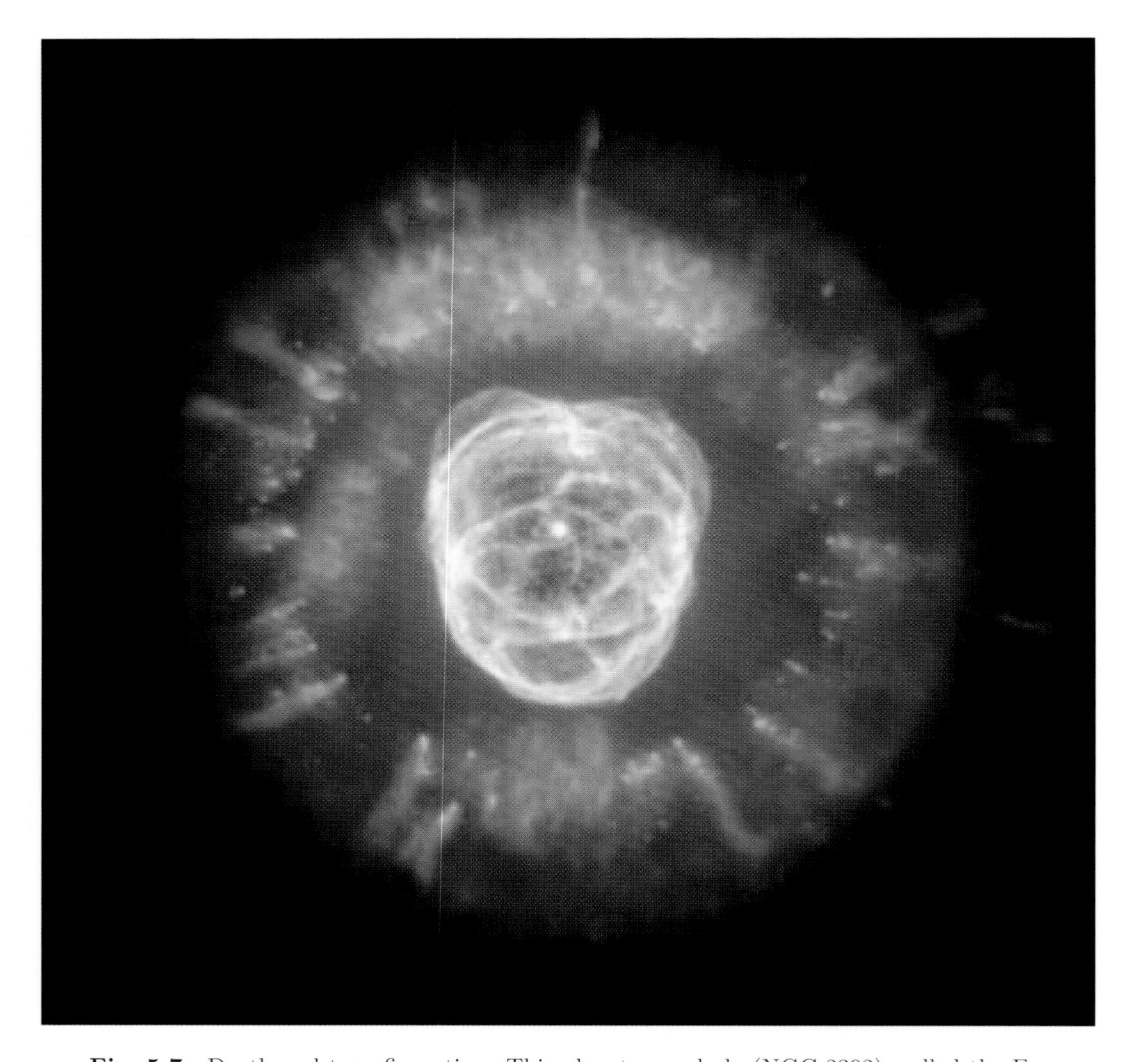

Fig. 5.7 Death and transfiguration: This planetary nebula (NGC 2392), called the Eskimo, began forming about 10,000 years ago when a star like our Sun began to die. In its place at the center a nascent white dwarf is forming. Neutron stars and black holes are formed in more cataclysmic events than this. The glowing remains of the star, first sighted by William Herschel in 1787, are being swept outward by winds from the hot white dwarf, the dense shell of which has a speed of 115,000 kilometers per hour. The winds at the edge have a speed of 1.5 million kilometer per hour, creating the elongated bubbles. Credit: NASA, ESA, Andrew Fruchter and the ERO team (STScI).

the universe is the Canada-France-Hawaii Telescope (CFHT) that operates the 3.6 meter telescope atop Mauna Kea, a dormant Hawaiian volcano rising 4,200 meters above the Pacific Ocean.

Fig. 5.8 Cerro Tololo Inter-American Observatory high on a rugged mountain top as seen from Cerro Pachón in north central Chile, April 2004. Credit: D. Isbell and National Optical Astronomy Observatory/Association of Universities for Research in Astronomy/National Science Foundation.

5.5. Questions

1. *How is the Sun weighed?*

   By the orbital radii and periods of the planets and an application of Kepler's laws (see page 19) $M_{\rm Sun} + M_{\rm Planet} = (2\pi)^2 R^3/(GP^2)$ where R and P are the average radius and period of the planetary orbit, and G is Newton's constant. Notice that the planet's mass can be ignored in comparison with the Sun's, so we have the "3/2-law", $M_{\rm Sun} = (2\pi)^2 R^3/(GP^2)$. That is to say, for every planet, R^3/P^2 (R is the average Sun-planet distance) has the same value. Indeed, P^2/R^3, where P is expressed in years and R in astronomical units (Au) has the value of 1 within one percent for every planet. For Earth, P is one year and R is one Au. Thus for Earth, P^2/R^3 is identically unity while for Mars it is 1.01.

2. *What type of galaxy is the Milky Way. Is it typical?*

   The Milky Way is a spiral arm galaxy and is a giant galaxy. This is the typical original form of galaxies. Other types are thought to have evolved from these, such as elliptical galaxies formed by a collision between spiral galaxies and their subsequent merger.

3. *How massive is the Milky Way and where is our planetary system located in it?*

   The mass of the Milky Way is about 10^{12} solar masses. Of this, about 10^{11} solar masses is in the form of luminous stars. Our solar system is located about two-thirds from the center.

4. *What accounts for the missing mass implied by the answer to the previous question?*

   Presumably it resides in non-luminous stars like neutron stars, white and brown dwarfs, but most likely most of it is in an as yet unidentified form of *dark* matter.

Chapter 6

Spacetime, Relativity, and Superdense Matter

Scarcely anyone who comprehends this theory can escape its magic.

A. Einstein

I was sitting in a chair in the patent office at Bern when all of a sudden a thought occurred to me: "If a person falls freely he will not feel his own weight." I was startled. This simple thought made a deep impression on me. It impelled me toward a theory of gravitation.

A. Einstein, 1907

6.1. Einstein: Spacetime and Relativity

"Absolute space... [and time] always remains similar and unmovable" (Isaac Newton in *The Principia*). Newton's laws of mechanics and gravity took on their simple form as he wrote them, only if space and time had a meaning independent of anything or anyone. For two hundred years his theory met with one success after another.

But by the early 1900s problems began to stir in a few minds. Maxwell's electromagnetic theory described many verifiable phenomena. It also predicted that the velocity of light is a constant of nature. This statement is so pregnant with meaning that it deserves amplification by contrast with Newton's notions of absolute space and time and our personal limited experience of them.

If a train is pursued by a car on a parallel road, the velocity *between* train and car is reduced. *Not so for light.* If a person were to try to follow a light beam (or photon of light) no matter how fast he went, he could not reduce the velocity with which it receded from him. This is the meaning of "the velocity of light is a constant of nature."

The seed of this profound notion had occurred to Albert Einstein (1879–1955) when he was only 16 years old, and it germinated in his mind until

Fig. 6.1 Albert Einstein in middle years, probably in Princeton, discoverer of the Special and General Theory of Relativity. The latter is the theory of gravity and is the foundation for our ability to trace the evolution of the universe from a small fraction of a second to its possible futures. ©Lotte Jacobi Collection.

finally in 1905 he gave it precise meaning. His original question, as he recalled it years later, was: "If I pursued a light wave at its own speed, I would see the waves standing still; but this cannot be — a wave standing still." His resolution: a material body cannot achieve the speed of light.

Ten years later, Einstein solved the dilemma of his youth posed by the independence of light velocity on the speed of the observer — and went further — by realizing that the velocity of light *is* indeed constant *but* space and time are *not* the absolutes as we think or that Newton before him thought. Indeed space and time have no separate existence from each other; rather, they are intertwined. Thus was born the Special Theory of Relativity (1905).

It was actually Minkowski who enunciated the fusion of space and time very clearly shortly after Einstein's theory was published. At a conference in Germany, Minkowski opened his lecture with these words,

> *The views of space and time which I wish to lay before you have sprung from the soil of experimental physics, and therein lies their strength. They are radical. Henceforth space by itself, and time by itself, are doomed to fade away into mere shadows, and only a kind of union of the two will preserve an independent reality.*
>
> Hermann Minkowski, *Lecture Sept. 21, 1908*

Thus space and time are fused into the single entity *spacetime*. This is much more than a semantic statement, though it may read like one. For the reader who is comfortable with mathematical formulae in stating physical concepts, refer to Box 15.

Einstein's new laws found almost immediate acceptance and they are daily confirmed as a matter of routine in the operation of the world's large particle accelerators. Our everyday experience is not contradicted by the fusion of space and time. For it is only when relative velocities are near the speed of light that differences between relative spacetime and our notion of absolute space and time show up. The pion, an important subatomic particle that is largely responsible for the nuclear force, lives for only 1/100 millionth of a second if it is measured by a person at rest *with respect to* the pion. Yet its lifetime is stretched almost infinitely with respect to the same person, if the pion is moving at a high velocity — close to the speed of light. This is a fact that is used routinely as a tool for experimentation. The connection that Einstein found that relates time t as measured by an observer moving with the pion whose velocity in the laboratory is v, and time T as measured by a stationary observer in the laboratory is

$$T = t/\sqrt{1-(v/c)^2}\,.$$

Translating this equation, we note that when v is close to the speed c of light then the denominator is close to zero so that the right side is therefore large. Recall that Einstein as a boy of 16 intuited that a material object cannot attain the speed, c. Of course the situation is entirely reversible; the motion is relative. Each observer sees time running slower as measured on his own stationary clock as compared to the other moving clock. The converse holds for distances as measured by the two observers, i.e.

$$L = l\sqrt{1-(v/c)^2}\,,$$

where as before capitals are measurements made by the stationary observer in the laboratory.

Still he was not satisfied with his theory. Special Relativity gave a favored status to situations in which relative speeds were constant as in the example of the train and the car. He could see no reason for such a favored status. Gravity provided the key. He knew that his new theory could not describe gravity because gravity causes acceleration, but his theory dealt with constant speeds. His theory made evident the fact that Newton's law of gravity could not be correct for bodies at high velocities, as in the example above of the stretched lifetime of a fast pion as observed by a stationary observer. In Newton's theory the mass of a body remained unchanged whatever its motion. And time too, flowed like a river — the same river — for everyone, and space, likewise, was absolute. Also, the distance between two stationary objects would be measured to be the same by all observers, no matter their motion, and similarly for the interval between two events.

Another serious problem with Newton's theory is the so-called "action at a distance." Einstein discovered that no signal could travel at a speed that is greater than that of light. Yet in Newton's theory, the Sun guided the motion of the Earth and other planets instantaneously. How else would they be guided in their motion from one moment to another? How, in other words, did a planet know where to move in the next instant if the signal were not assumed to travel instantaneously from the Sun? In Einstein's theory the answer is provided by the curvature of space produced by all matter and energy. As a *rough* picture, imagine a massive body sitting on an elastic sheet in which it makes a deep indentation. A lighter body produces its own indentation, but in the presence of the deep one, the lighter body will tend to fall toward the large one. But if its motion is tangential then it will move around the larger body on a circle as depicted in Figure 6.2. No instantaneous signal between the bodies is needed to tell them where to move.

The way to a theory of gravity, often called General Relativity, was beset by many difficulties, but years later in his Princeton lectures he was able to say precisely what bothered him about the concept of spacetime held at that time in a way that all scientists could relate to: "...it is contrary to the mode of thinking in science to conceive of a thing (spacetime) which

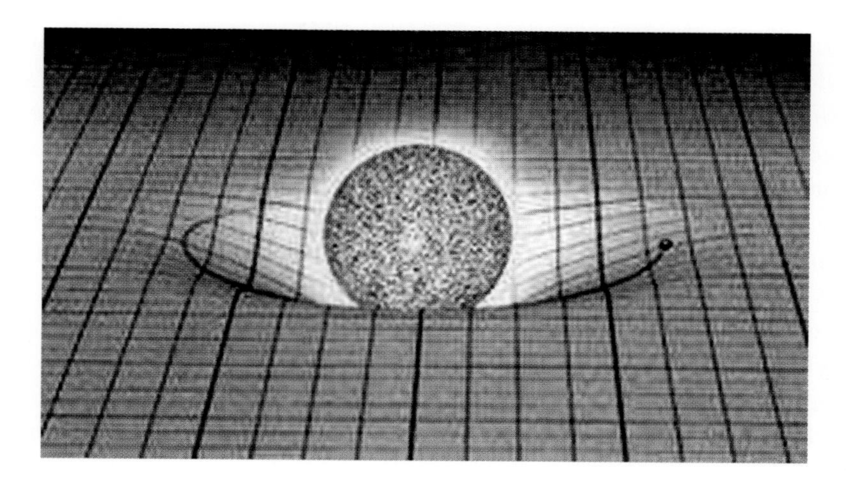

Fig. 6.2 The orbit of a light body around a heavy one is depicted here in two dimensions: it moves in the indentation in spacetime produced by the heavier. The tangential velocity of the small body is what keeps it from falling toward the larger. Instead it circles, with a radius depending on its velocity. If the picture were drawn accurately, the Earth, also, would make an indentation in spacetime, only smaller than that made by the Sun. This gives only a rough idea of how matter — and energy of all forms, including motion — warps spacetime.

acts, but cannot be acted upon."[a] This is Newton's mechanical law — to every action there is an equal and opposite reaction. But unlike Newton, Einstein asserted the generality of the law in every respect. Therefore if the arrangement of matter and its motion shape spacetime, spacetime in turn effects matter and its motion.

It was a great struggle to give exact meaning to this idea (1907–1915). At first he did not even know there was a mathematics suitable for his purpose, that certain mathematicians had formulated a geometry of curved space, Riemann being foremost among them. Nevertheless, within a few short years, and a few false starts, he was able to write his law of gravity, a law that gave precise meaning to his notion that spacetime must be acted upon by mass-energy, just as mass-energy is shaped and moved by spacetime. In symbolic form, his theory of gravity (1915) can be written as a single equation

$$\mathbf{G} = -8\pi\mathbf{T}\,.$$

The symbol $\mathbf{G}$ is known as Einstein's curvature tensor; it describes the geometry of spacetime — its curvature. The term on the right, $\mathbf{T}$, describes

[a]Quotation from the third chapter of his marvelous little book *The Meaning of Relativity.*

matter and its motion in a particular way. The equality between the left and right sides of the equation amounts to saying that *matter and its motion tell spacetime how to curve, and in turn, spacetime tells matter how to move and arrange itself.* Spacetime is no longer a fixed arena as in Newtonian physics, in which things happen, but which itself is unaffected by what happens. Rather, it is an integral part of nature, responding to and shaping what is happening. This was a radical departure of any other perception of our world.

Later, Einstein introduced the cosmological constant, Λ, into his equations when he learned that without it, the universe would eventually collapse, or expand in a coasting fashion forever. At that time it was believed that the universe was static. So did Newton before him. For that reason Newton believed that the universe was infinite for otherwise, he reasoned, gravity would cause it to collapse. Einstein chose a value of the cosmological constant that would cause the universe to remain unchanged, neither expanding or contracting. When he learned later from Hubble that the universe is in fact expanding, he renounced it. Interestingly, it remained an object of interest to some scientists. In our own day, it has been discovered that the cosmic expansion is accelerating, and that Λ could represent a constant energy density — called dark energy — that pervades the universe, and drives it eventually into *accelerating* expansion; this is the stage at which the universe arrived several billion years ago. Einstein gave the cosmological constant its name because it is small and acts only on the scale of the universe, not on individual stars or galaxies.

In Einstein's theory of gravity, popularly known as *General Relativity*, the matter and motion of Sun and Earth warp spacetime, and the Earth moves in the groves of this warped spacetime (Figure 6.2). Because matter warps spacetime, Einstein predicted that light from a distant star, in passing near the Sun to us, would be deflected by twice the amount predicted by Newton's theory. He was right, as discovered by Eddington, who made an expedition to Principe off the coast of Africa in 1919 for the express purpose of making the measurement.

Einstein's theory of gravity transforms seamlessly into Newton's theory when gravity is weak. However, Einstein's theory has a remarkable property; when gravity is weak the force law is the inverse square of the distance — precisely. It can be nothing else in his theory. In Newton's theory, the inverse square was postulated because then the theory worked well, and perhaps because Newton was partial to integers. It could have been the inverse of distance raised to the 1.99 power, for example, and the solar system dynamics would have worked almost the same, and within the normal errors and perturbations from undetected objects like minor planets or meteor belts. There was nothing about Newton's theory that absolutely demanded the *integer* power 2. In Einstein's gravity it could be nothing else.

Neither Einstein, nor anyone else at the time could foresee all the wonders that would flow from such a seemingly simple equation as the one written above. Of course it was understood, even from the beginning, that the equation was much more complex than it appeared — that in fact it represents many equations referred to by mathematicians as simultaneous — that none of the equations has a precise meaning by itself but they have meaning only in their entirety. They have *another* property that mathematicians refer to as non-linearity. In a simple way of describing non-linearity we describe its opposite. Addition of two numbers is a linear operation. The operation always produces another number. There are no surprises. But non-linearity in mathematics can lead to results that, while determined by the equations, are often totally unforeseen.

One of the results that was not immediately foreseen is the existence of a type of star, a dead star such as a white dwarf or neutron star. These stars have evolved to the endpoint of their lives — neither produces energy by atomic or nuclear process and are quite inert. For such stars there is a mass limit — a mass above which no star of the type could have a larger value. Subrahmanyan Chandrasekhar discovered this property for the class of stars known as white dwarfs. (Neutron stars were yet to be discovered by Jocelyn Bell.) Eddington, who had made many marvelous discoveries of his own, refused to accept Chandrasekhar's idea. He probably saw in it another prediction, which he considered to be "absurd" — black holes. Eddington was wrong in this respect, because continued gravitational collapse to a black hole is now understood to become inevitable for stars too small for their mass. Neither Einstein nor others of his time suspected this. However, as we have seen, the Reverend John Michell, 200 hundred years earlier, had surmised the possible existence of black holes and the means of detection, now used in our day (page 92).

It remained for Oppenheimer (Figure 6.3) and Volkoff to discover in 1939 the true nature of gravity's grasp on the mass of dead relativistic stars (Figure 6.4). They studied the prediction of Einstein's equations for a spherical star that could produce no energy — because — in the language of physics — its matter was in a degenerate state — the state of lowest energy at whatever density of the matter. They invented a simple model of neutron stars almost thirty years before their discovery. And they showed that their

Fig. 6.3 J. Robert Oppenheimer, a leading American theoretical physicist, greatly admired by several generations of physicists for his physics, his integrity, and his loyalty to the nation. He was director of the Los Alamos Laboratory during the development of the atomic bomb. The photograph was taken by Ulli Steltzer.

mass is subject to an upper limit above which a neutron star would sink into a black hole. The limit exists simply because the world is made that way.

Oppenheimer and a student named Snyder took a step further in their study of the "absurd" notion implied by a mass limit of dead stars — continued gravitational collapse — forever — into a black hole if the mass exceeded the limit. Even though following as a logical consequence of Einstein's equations, the possible existence of such objects as black holes was not at first accepted, even by such an expert and visionary as John Wheeler. At a 1958 conference in Brussels, he concluded a brilliant lecture on the theorems and discoveries of his group at Princeton concerning the two families of dead stars, white dwarfs and neutron stars, by arguing that some physical process would intervene to prevent continued collapse. Oppenheimer politely demurred — "I believe ... in ... General Relativity."

Fig. 6.4 An artist's conception of the distortion of space in the vicinity of a rotating black hole. Similarly, for any rotating body except that the funnel end is rounded and shorter the larger the mass. Credit: Sky and Telescope.

Not long afterward, Wheeler accepted the logical imperative, and coined the name "black hole."

In a curious footnote to history, the notion of a star "whose light could not arrive at us", as we have learned (pages 25 and 92), was conceived by a British cleric and geologist, the Reverend John Michell in a letter to John Cavendish in 1783.

He reasoned that Newton's law of gravity would act on light, just as it does on mass. In this case he calculated the condition that light could not escape from stars having a certain relation between their mass and size. Curiously the derived relation (in classical theory) is precisely the one that follows from Einstein's theory, known as the Schwarzschild radius of a black hole, although Michell arrived at it by incorrect reasoning; he thought that light would be slowed down in its flight from a star, like a stone thrown into the sky. He did not know what Einstein was to discover two centuries later, that light has a constant velocity. What is essential, however, is that the energy of the photons of light would be reduced in leaving the vicinity of a black hole. And none could escape from within.

Michell also devised the only means we have of detecting a black hole — "still we might infer the existence of the central object by its effect on the motion of a luminous companion." The world was not ready for these ideas. Michell's prescient idea was uncovered long after the modern theory of gravity and black holes had been discovered.

A neutron star, at the limit of the mass for a dead star, hovers on the verge of collapse into a black hole. If mass were somehow added say by an asteroid falling upon it, it would begin its never-ending collapse from which not even light can escape. At least, this would be so if it were not for quantum mechanics. As Stephen Hawking discovered (1974), black holes have a peculiar property akin to temperature that depends on their mass. Massive black holes have a very low temperature and they will not evaporate except on a very long time-scale that typically exceeds the present age of the universe. However, lighter black holes whose temperature is greater than that of the present ambient temperature of the universe, about 3 degrees kelvin, will evaporate (see page 125).

There are other peculiar properties of black holes. If, from outside, one were to watch an object fall into a black hole, the object would appear to travel ever more slowly and take forever to fall within. But for an observer falling in, his watch would reveal nothing so peculiar — his own time would flow without change as far as he was concerned. What happens to the flow of time in the strong gravitational field near a black hole is only a more extreme example of what has been actually measured by careful experimentation on Earth: time flows differently in different gravitational fields!

Because of their elusive properties, it is difficult to say with absolute certainty that any black hole has been discovered. Still, there are "candidates." Just as the Reverend John Michell suggested, there are unseen objects whose presence can be inferred from their effects on a luminous stellar companion — a visible star in orbit with no visible partner. In some cases the companion could be a neutron star that is not beaming in our direction. However, in other cases the inferred mass of the unseen star is too large for a neutron star — the mass lies above the mass limit set by relativity and the nature of matter. Black holes are the only known objects having a sound basis in physics that could be the invisible companions. This may remain the strongest type of evidence for the existence of black holes.

6.2. Superdense Matter

Neutron stars were first conceived of as being just that, stars made of closely packed neutrons. Oppenheimer, Volkoff, and Tolman estimated the approximate mass range of such stars. They discovered that dead relativistic stars have an upper bound for their mass that no neutron star can exceed. Their estimate — based on a theoretical model in which the dense star contained only closely packed neutrons — was three-quarters of the mass of the Sun. This pioneering calculation did not include the well-known repulsion between nucleons when they are close together, but this does not detract from their achievement in taking seriously the problem of relativistic stars and solving Einstein's equations for a model star. The nuclear repulsion at short-range raises the maximum possible mass of neutron stars to a value close to 2 solar masses because the repulsion introduces a resistance to the pull of gravity. On the other hand, the fact that the nucleon has excited states may counteract the repulsion. So it remains an open question what the limit is. The masses of a few neutron stars have in fact been determined, and the largest of these places a lower limit on the mass. It could be higher, but not lower. Within the uncertainties in the observed masses, present data places the upper limit between 1.44 and 2 solar masses — not a very precise limit, but useful nonetheless. For if a neutron star had a greater mass than the limit, it would sink inexorably into a black hole.

Still there are many uncertainties regarding the nature of superdense matter. The immense densities found in the interior of neutron stars is nothing like we will ever know on Earth except possibly in the fleeting moment following the collision of very fast nuclei in the largest accelerators in the world. Nevertheless, we can use established laws and principles to guide us in the investigation of possible states of superdense matter in the interior of neutron stars, being always alert to what observable effect they might have. In this way we are offered the possibility of discovering properties of superdense matter that are not known from laboratory experiments — matter that can never exist on earth. And in so doing, we can elucidate and extend the laws of nature and the range of known phenomena.

Thus from the laws of physics including Einstein's theory of gravity, and the well substantiated theory of the underlying quark structure of nucleons known as "quantum chromodynamics" or simply QCD, we are permitted to investigate a range of possible phenomena that may occur in superdense matter. Only in its simplest conception can neutron star matter be thought of as a uniform medium of closely spaced nucleons. Phase transitions may occur at successively higher densities that fundamentally change the nature of dense matter. Some phase transitions that we are familiar with are quite simple such as the transition of water to ice or steam. In others, particularly in substances that have a richer composition to begin with, phase transitions can introduce quite unexpected new structure to the substance — a geometrical lattice of quite marvelous nature. Neutron star matter belongs to the latter category [Glendenning (2000)].

From the discovery of pulsars that rotate up to 600 times a second, such as the one discovered in 1982 by Donald Backer (or the recent (2006) discovery of an even faster one, rotating at 716 times second, which has been found in a globular cluster of stars called *Terazan 5*), we can estimate the range of densities of matter that occur in neutron stars. The estimate is made by comparing the centrifugal force on a particle at the surface of a star rotating at 600 times a second with the gravitational attraction of the rest of the star. Gravity must win or the star would fly apart at the equator. By such a comparison we learn that the *average* density is a little more than the density of atomic nuclei. Since the density is largest at the center of a star and decreases to a small value at the surface, the central density must be a few times nuclear density. From laboratory experiments we know the size of nucleons. At the central density of neutron stars, the nucleons would overlap. Therefore, just as nuclei themselves dissolve at high density into matter composed of the nuclear constituents — the nucleons — so too we expect that nucleons will dissolve into their constituents — the quarks.

Quarks have a peculiar property compared to the constituents of matter at all other known scales. Molecules can be taken apart and their constituents, the atoms can be studied. Atoms can likewise be taken apart and the central nucleus studied. The nucleus too can be separated into its nucleons, and reactions among them have been investigated in all their complexity. But it is apparently a law of nature that the quarks of an individual nucleon cannot be removed and isolated for study. Nevertheless, it is a prediction of the theory that if nucleons are crushed to very close proximity, as at the center of a neutron star, the quarks will lose their association with their previous hosts. The individual quarks still cannot be removed from such a crushed dense state of matter. But this exotic state of matter in bulk at high densities is among the most fascinating we can imagine.

Such a crushed state of matter is called *quark matter.* It must have pervaded the universe in its very earliest instants — before expanding from its hot and dense beginning — cooling as it did so and making space for three quarks to combine to form individual neutrons and protons. The neutrons and protons in turn combined to form atomic nuclei, which make virtually all the mass of our bodies. The combination of quarks into neutrons and protons occurred at a time of 1/100 thousandth of a second when the temperature had fallen to a thousand billion (10^{12}) °C. The combination of neutrons and protons into the light nuclei — hydrogen, helium, up to lithium took place in the first few minutes.

Those elements are referred to as the *primordial elements* because heavier elements were not formed until the first stars evolved from great clouds of these elements at about 200,000 years. These clouds are often referred to as molecular clouds, because by that time, electrons had combined with the nuclei of the above elements to form atoms and some of these combined to form molecules. This sequence of events occurred as the universe cooled.

6.3. Spinning Neutron Stars

Such a state of matter — quark matter — that existed in the fiery furnace at the beginning of time may exist now in the deep interior of neutron stars. They are very dense by earthly standards — nor is there room enough for individual nucleons to exist in their cores. The nucleons would be crushed, laying bare their constituent quarks. How would it be possible to learn if indeed this primordial state of matter has been recreated in some neutron stars?

It is beautifully simple to understand how we here on Earth may learn what is happening in the heart of a neutron star. The same law of physics used by ice skaters when they spin up from a slow twirl to a dazzling whirl provides the signal. The law is the conservation of angular momentum, or spin as we will call it for brevity.

An analogous thing will happen to a rotating neutron star if its center converts to quark matter. Imagine a rapidly rotating star. It will be slightly flattened because of the centrifugal force, somewhat like a pancake. A huge magnet field is embedded in the star, along which radio waves are beamed like the beacon of a lighthouse. The radio waves and other radiation carry off rotational energy and therefore the star very slowly loses spin. The loss of spin by the star is not a violation of the conservation law of angular momentum, because what the star loses, the outside world gains. Part of this gain is what powers the very radio antennae by which we can 'see' the spinning pulsar.

As the star slowly loses its spin, the centrifugal force decreases and the star becomes less pancake shaped — becomes more spherical. Its mass becomes more centrally concentrated and its central density increases. When the density has increased to the critical value such that individual nucleons overlap, their quarks break free into the state of quark matter — like the one described above that pervaded the early universe. This state is far less resistant to compression than the nuclear matter that formerly occupied that region of the star. Therefore, the rest of the star, weighing down on it, compresses and reduces the size of the star. This process continues for possibly 10 million years. However, as the spinning star becomes smaller, it must spin faster to conserve angular momentum that is not carried away fast enough by the very inefficient way in which the star radiates angular momentum through its radio waves that form the beacon described above.

Just like the ice skater who concentrates his weight closer to his center and spins faster as a result, the star, by converting nuclear matter to quark matter, concentrates its weight closer to its center and spins faster so as to conserve angular momentum. Just as the ice skater's performance is quite stunning, so too such a performance by a star would be quite stunning. Stunning, because the natural state of affairs when a star or anything else loses or radiates spin is that it spins slower — not faster. In this manner, we may hear from deep within the heart of a neutron star the signal of the state in which matter existed in the very early universe.

6.4. Questions

1. *Using the information in Box 17 how long will it take for a 10 $M_\odot$ black hole to evaporate.*
   2.11×10^{70} years

2. *Express the evaporation time τ and temperature T of a black hole in terms of a solar mass.*
$$\tau = \frac{[M_\odot c^2]^3}{3Kc^4}\left[\frac{M}{M_\odot}\right]^3, \qquad T = \frac{\hbar c^3}{8\pi G k M_\odot}\frac{M_\odot}{M}$$

3. *How massive would a black hole that survived to the present day (i.e. 15 billion years) be?*
   $M/M_\odot = 8.93 \times 10^{-20}$. Note that the Sun's mass is 2×10^{33} grams.

4. *How long will it take for a 1 gram black hole to evaporate?*
   8.3×10^{-24} seconds

6.5. Boxes 13–18

13 Primordial Black Holes

It is believed that at very early times there were many primordial black holes. Only the heaviest would have evaded evaporation to the present time. Perhaps they are the seeds of the giant black holes at the centers of galaxies or even the seeds about which primordial clouds condensed.

14 Einstein's Equations

Einstein's equation can be written,

$$\mathbf{G} = -8\pi\mathbf{T} + \Lambda\mathbf{g}\,.$$

Here we have now added, as Einstein did, what he called the cosmological term, Λ. The symbols written in capitals are called tensors and are of second rank; they are analogous to vectors but more complicated. $\mathbf{G}$ on the left is known as Einstein's curvature tensor; it describes the geometry of spacetime — its curvature. The term on the right, $\mathbf{T}$, describes in a particular way matter *and* its motion. The equality between the left and right sides of the equation amounts to saying that matter and its motion tell spacetime how to curve, and in turn, spacetime tells matter how to move and arrange itself. $\mathbf{g}$ is called the *metric tensor* and is the mathematical way in which the curvature of spacetime is described: the collection of components in $\mathbf{g}$ take the place in General Relativity of the Newtonian potential V, and therefore they are sometimes called the gravitational fields. The Einstein curvature tensor, $\mathbf{G}$ is made up in a particular way by the elements of $\mathbf{g}$. The cosmological constant is denoted by Λ.

15 Invariant Interval in Relativity

The distance dl between two events, P_1 and P_2 with coordinates (x_1, y_1, z_1) and (x_2, y_2, z_2) is

$$dl^2 = dx^2 + dy^2 + dz^2 ,$$

in Euclidean geometry, where the notation $dx = x_2 - x_1$, etc. is used. This is Euclid's geometry, which holds on an ideal flat plane, and in practice, in the presence of a gravity, provided space curvature is small over the dimensions considered. Any two experimenters who make their measurements carefully will agree on the result of the measurement of distance dl. The time at which the events took place plays no role in such a measurement, in Euclid's geometry.

In the Special Theory of Relativity, time also enters the measurement. The effects are too small to perceive in everyday life. In general, let t_1 denote the time at which an event or measurement at P_1 occurs and similarly for '2'. Then it is a particular combination of the time interval between the events and the distance between them that remains unchanged, *no matter what particular (Lorentz) coordinate system* has been used to define the positions and measure the times at which they occurred;

$$d\tau^2 = dt^2 - (dx^2 + dy^2 + dz^2)/c^2 .$$

$d\tau$ is a definition of the *proper time*, or *invariant interval* between the events.

In the Euclidean case the spatial distance dl between the two points P_1 and P_2 is what remains unchanged or invariant. In Special Relativity the invariant is the combination of space and time intervals, namely the invariant interval $d\tau$.

In the particular case that a *light* signal is sent from P_1 to P_2 then $d\tau$ is zero, and the above equation is a statement of the constancy of light speed where a light signal is sent from P_1 to P_2 and arrives in the time interval $dt = t_2 - t_1$.

16 Time Dilation

Let there be two identical clocks at rest at $x = 0$. The clocks tick at equal intervals dt. The proper time in this frame is

$$d\tau = \sqrt{dt^2 - (dx/c)^2} = dt\,.$$

An observer takes one of the clocks and moves away along the x-axis with velocity v. He sees the first clock at $dx' = -v\,dt$. The proper time expressed in his frame is

$$d\tau' = \sqrt{dt'^2 - (dx'/c)^2} = dt'\sqrt{1-(v/c)^2}\,.$$

But the proper times are the same. Consequently

$$dt' = dt/\sqrt{1-(v/c)^2}\,.$$

This expresses the dilation of time as measured by observers in uniform relative motion. The converse holds for length measurements.

17 Black Hole Evaporation

The Hawking temperature T of a Schwarzschild black hole with mass M is given by

$$T = (\hbar c^3)/(8\pi GkM)\,,$$

where k is the Boltzmann constant. The emission of this energy results in an energy decrease of the black hole, and thus a loss in its mass. What period of time τ will it take for a black hole of mass M to evaporate completely? The Schwarzschild radius of the black hole is $R = 2GM/c^2$ and it thus has an area of $A = 4\pi R^2$ or

$$A = 16\pi G^2 M^2/c^4\,.$$

Hawking radiation would have a power P related to the hole's area, A, and its temperature, T, by the black-body power law, $P = \sigma AT^4$ (Steffan–Boltzmann constant σ) which, gathering the above results yields

$$P = K/M^2\,,$$
$$K = (\sigma\hbar^4 c^8)/(256\pi^3 G^2 k^4) = 3.55 \times 10^{32}\,\mathrm{W}\cdot\mathrm{kg}^2\,.$$

But the power of the Hawking radiation is the rate of energy loss of the hole; therefore we can write $P = -dE/dt$. Since the total energy E of the hole is related to its mass M by $E = Mc^2$, we can rewrite the power as $P = -c^2 dM/dt$, so that

$$-c^2 dM/dt = K/M^2\,.$$

Therefore,

$$M^2 dM = -[K/c^2]dt\,.$$

Integrate over M from the initial mass M to 0 and over t from zero to τ, to find the evaporation time,

$$\tau = [c^2/(3K)]M^3 \approx 2.11\,(M/M_\odot)^3 \times 10^{67}\ \mathrm{yr}\,.$$

The mass of a black hole that will take the life of the universe of 15 billion years to evaporate is,

$$M/M_\odot = 8.9 \times 10^{-20}\,.$$

18 Evaporation in the Environment of the Background Radiation

Keep in mind that real black holes are immersed in a universe in which radiation is present, namely the cosmic background radiation (CMBR). The universal temperature in the present epoch is $T_{\rm CBR} = 2.7$ K. Because the Hawking temperature is inverse to the black hole mass, there is a small but limiting mass to which the black hole evaporates, which we call m. It can be related to the present background temperature by the formula developed in the previous box.

$$m = (\hbar c^3)/(8\pi G k T_{\rm CMBR})\,.$$

The time for a black hole of mass M to evaporate to this presently limiting mass is

$$\tau = [c^2/3K](M - m)^3\,.$$

Chapter 7

Origins

For I am yearning to visit the limits of the all-nurturing Earth, and Oceans, from whom the gods are sprung ... [Hera to Aphrodite]

Homer, *Iliad 14.201*

7.1. Formation of the Solar System

Every 12 years Jupiter returns to the same position in the sky; every 370 days it disappears in the fire of the Sun in the evening to the west, 30 days later it reappears in the morning to the east ...

Gan De, *Chinese astronomer, born about 400 B.C.*

The solar system — our Sun, its planets, and their moons — lie within a great galaxy called the Milky Way, which we can see on a clear night as

Fig. 7.1 Io is disguised as a young heifer by Jupiter. Juno (Hera), queen of the gods, one day perceived it suddenly grow dark, and immediately suspected that her husband, Jupiter, had raised a cloud to hide his philandering with the fair young Io, daughter of the river god Inachus. She brushed away the cloud, and saw her husband on the banks of a glassy river, with a beautiful heifer standing near him. Juno suspected the heifer's form concealed the fair nymph. Indeed, it was so.

Fig. 7.2 A view of our galaxy, the Milky Way, taken from Earth and looking toward the center. Our solar system is about two-thirds out from the central bulge. Courtesy: NASA.

a pale band of light created by billions of stars.[a] The stars in the Milky Way are made in part from the remains of very large, ancient and short-lived stars that had formed about 200,000 years *after the beginning.* Their debris dispersed and wandered through the universe for billions of years, eventually joining the primeval gases of hydrogen and helium and debris of more recent defunct stars to form enormous clouds. The clouds began to fragment, taking up to a hundred million years to collapse into rotating *nebulous* galaxies, which in turn fragmented into stars, the light of which we see as the Milky Way, Figure 7.2.

Rotation caused such a nebulous cloud to spread into a large thin disk with a central bulge until a balance was reached between the forces exerted by gravity and the rotating centrifuge. Immanuel Kant, the German philosopher, and Pierre Simon de Laplace (Figure 7.3), the French astronomer and mathematician and a student of d'Alembert, speculated on this *nebular hypothesis.* Laplace presented his famous hypothesis in 1796 in *Exposition du systeme du monde*, written in five volumes. In the last he viewed the solar system as originating from the contraction and cooling of a large and slowly rotating cloud of incandescent gas stretched thin by its rotation.

[a]Unfortunately for city dwellers the view is obscured by light from street lamps and houses, reflected from pollutants, particles of dust and droplets of water in the atmosphere.

Fig. 7.3 Emmanuel Kant (1724–1804), German philosopher, and Pierre Simon de Laplace (1749–1827), French philosopher and astronomer, who each conceived the notion of the origin of the solar system in a nebulous cloud.

Indeed, scientists today believe that the solar system was formed when such a disk of gas and dust was disturbed, perhaps by the explosion of a nearby star — a supernova. This explosion created in the gas, waves of alternating higher and lower density. Under special circumstances, gravity will cause the collapse of a region of high density to form a solar nebula (Boxes 11 and 12). The collapse of a portion of a solar-nebula-sized molecular cloud, eventually to form the solar system occurred in a relatively short time of a few million years. Evidence for the speed of this process comes from the technique used in archeology — radioactive dating. However, for astrophysical purposes, the radioactive isotopes that are useful must be long-lived (see box note on page 153).

As collapse progressed, which because of the size took roughly 10 million years, the cloud grew hotter and denser, especially in the center where heat was trapped. Particles began to stick together and form hot molten clumps. They grew bigger as other clumps stuck to them so that gravity became effective in attracting clumps of matter at greater distance to form planetesimals. Eventually planetesimals were caught up by gravity to form planets and moons.

The aggregation of interstellar dust (less than about 0.1 m in diameter) into increasingly large bodies occurred by random collisions. Eventually kilometer-sized planetesimals formed, large enough for gravity to speed the growth, culminating in the formation of asteroids and planets, which took place over a time interval of some 8 million years following the formation of the interstellar cloud.

7.3. The Sun

Sol comprises 98 percent of the solar system mass: Let us look now at its magical powers...

For all appearances, our Sun is a quiescent object, peaceful, warm and bright. So much for appearances. The Sun is a fiery cauldron, from its center through every layer, to its flaming surface. Even beyond; it radiates charged particles into space that set off the Northern Lights, known by the Latin name as the *aurora borealis*. The magnetic field of Earth usually protects us from this harmful radiation — sometimes. On March 13, 1989 the Sun ejected billions of tons of flaming hydrogen gas toward the Earth. The impact, which reached Earth in three days, caused huge electrical currents to surge through power lines in Eastern Canada, transformers to smoke, relays to melt. The entire power grid of the province of Quebec was shut off.

The Sun has strong magnetic fields, not a single one with a fixed orientation as on Earth, but many, appearing and disappearing in many places because of the great internal turmoil caused by the transport of heat to the surface. Associated with the magnetic activity are sunspots which appear as dark spots on the surface of the Sun (Figures 7.6, 7.7). They typically last for several days, although very large ones may live for several weeks. Sunspots are magnetic regions on the Sun with magnetic field strengths thousands of times stronger than the Earth's magnetic field. Sunspots usually come in groups with two sets of spots. One set will have a north magnetic field while the other set will have a south magnetic field. The field is strongest in the darker parts of the sunspots — the umbra. The field is weaker and more horizontal in the lighter part — the penumbra.

Aside from the sunspot itself shown in Figure 7.6, the surrounding grain-like structure of the Sun's surface may seem puzzling inasmuch as the Sun's surface appears to us to be smooth. Actually this is not so; telescopic photos reveal a mottled surface. Because the Sun's interior temperature and density are not constant, heat flows from the inner hottest regions to the surface by

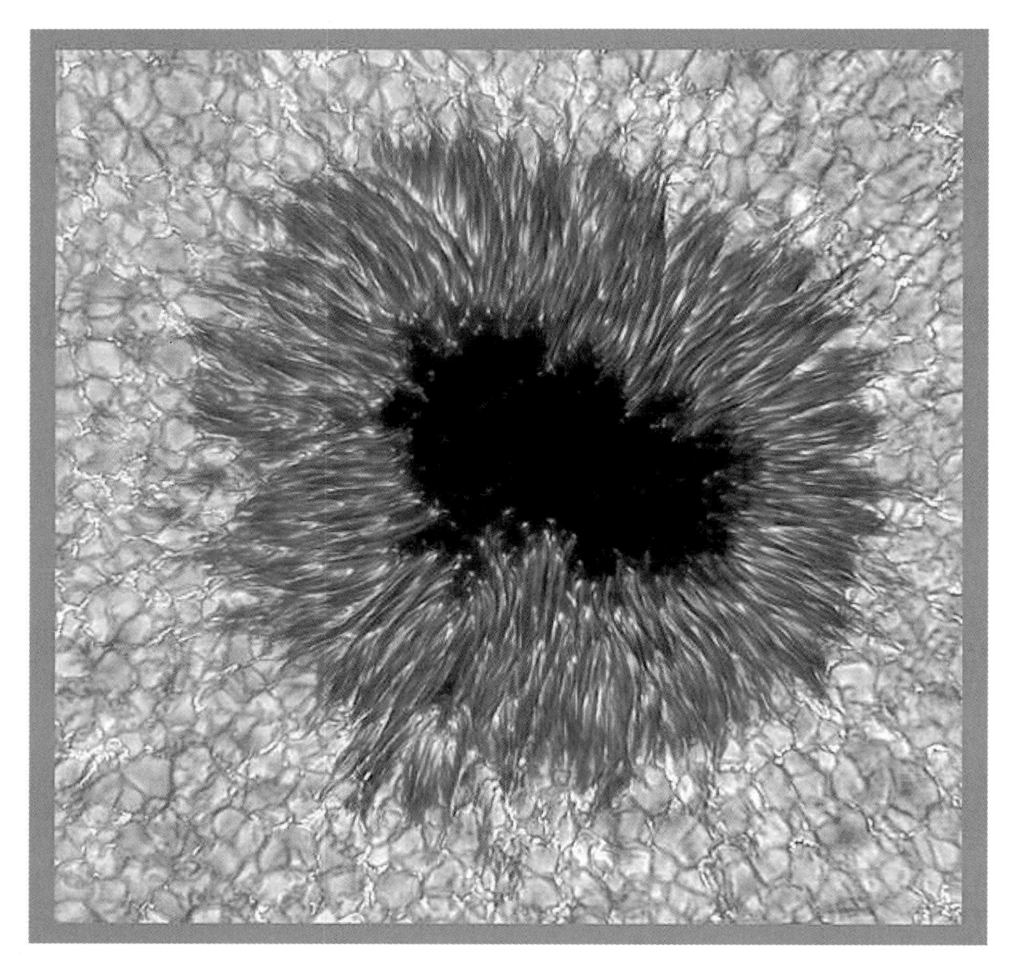

Fig. 7.6 High resolution image of a sunspot three times the size of the Earth. Image credit: Friedrich Woeger, KIS, and Chris Berst and Mark Komsa, NSO/AURA/NSF.

convective columns; the upward flowing gas columns can be seen in the figure as irregularly shaped 'nodules'. The downward flowing convection is of cooler denser gas through narrow channels, which surround the upward flowing columns and are seen in the figure by the narrow dark outline of the 'nodules'.

In 1610, shortly after viewing the Sun with his new telescope, Galileo Galilei made the first European observations of sunspots. Daily observations were started at the Zurich Observatory in 1749 and with the addition of other observatories continuous observations were obtained starting in 1849. Early records of sunspots indicate that the Sun went through a period of inactivity

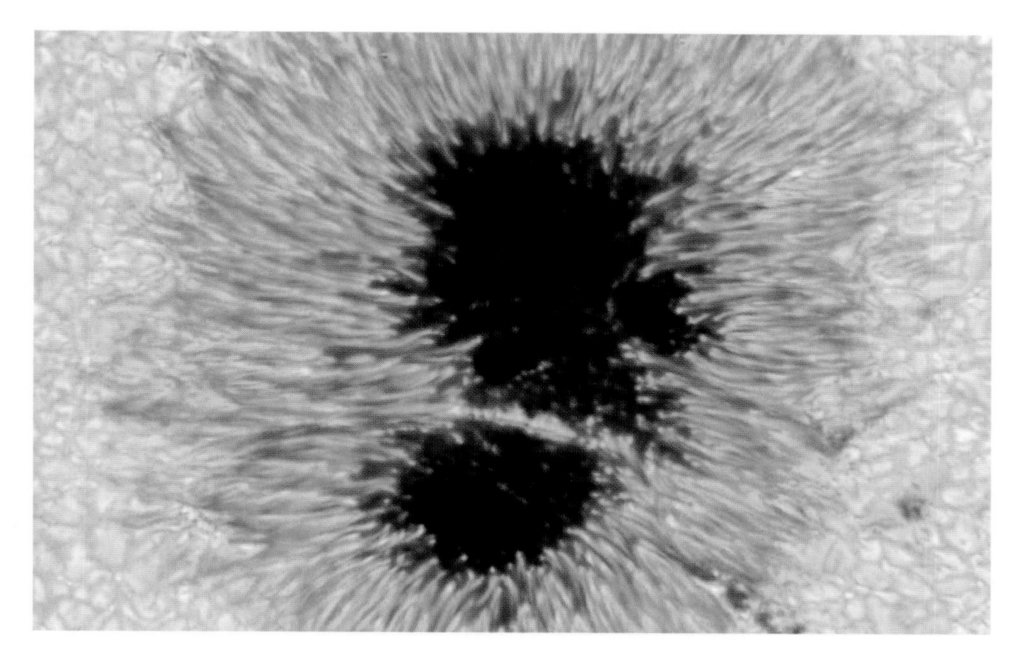

Fig. 7.7 A large sunspot observed on the Sun in September 2004. The entire Earth would fit into the area shown several times over. Sunspots appear dark because the strong magnetic field in the them suppresses the transport of energy through gas flow. In the central dark area of the sunspot (umbra) the magnetic field is perpendicular to the surface, whereas in the lighter colored periphery (penumbra) the magnetic field is largely horizontal to the surface. The image was captured by Vasily Zakharov with a one-meter solar telescope on the island of La Palma. The telescope is operated by the Institute for Solar Physics of the Royal Swedish Academy of Sciences. Image courtesy of Vasily Zakharov and the Max Planck Institute for Solar System Research.

in the late 17th century. Very few sunspots were seen on the Sun from about 1645 to 1715. The period of solar inactivity corresponds to a climatic period called the "Little Ice Age" when rivers that are normally ice-free froze and snow fields remained year-round even at lower altitudes.

But the history of sunspots has also been deciphered over a much longer time span by analyzing radioactive isotopes in trees that lived thousands of years ago. Solar activity over the last 11,400 years, back to the end of the last ice age on Earth, has now been reconstructed quantitatively by an international group of researchers (Germany, Finland, and Switzerland) sponsored by the Max Planck Institute for Solar System Research (Katlenburg-Lindau, Germany). It was found that the Sun has been more active during the past 60 years than in the last 8000 years. Based on a statistical study of earlier periods of increased solar activity, the researchers predict that the current level of high solar activity will probably continue only for a few more decades.

7.4. The Sun and Arthur Eddington

We are bits of stellar matter that got cold by accident, bits of a star gone wrong.

Sir Arthur Eddington (1882–1944)

There are some parallels between the lives of Sir Arthur Eddington (1882–1944) and Sir Isaac Newton (1642–1727). They were both born into families of modest means, their fathers died early, Eddington's when he was two and Newton's before he was born. Both men were recognized early by their teachers as being exceptional, and were assisted in their education by scholarships. Each achieved fame in his own time and was knighted by the crown. Neither was married or had children.

Arthur Eddington's[c] father taught at a Quaker training college in Lancashire. He died of typhoid in an epidemic that swept the country before his son was two years old. Eddington's mother, like her husband, was also from a Quaker family. On her husband's death she was left to bring up Arthur and his older sister with relatively little income. Arthur was educated at home before spending three years at a preparatory school, which was mainly for boarders. However, he did not board at the school, being one of a small number of day pupils. The school provided a good education within the limited resources available to it. Arthur excelled in mathematics and English literature. His progress through the school was rapid and he earned high distinction in mathematics.

Eddington was awarded a scholarship at the age of sixteen, but officially he was too young to enter university. It was a problem that was solved quickly, however, and did not cause him to delay entry to Owens College, Manchester. In his first year of study, Eddington took general subjects before spending the next three years studying mainly physics and mathematics. Eddington was greatly influenced by one of his mathematics teachers, Horace Lamb. His outstanding academic work allowed him to win a number of highly competitive scholarships.

At the age of 19, Eddington was awarded a Natural Science scholarship to study at Trinity College, Cambridge. At Trinity one of his teachers was the great mathematician E. T. Whittaker. After graduating, he began research in the Cavendish Laboratory on several physics topics but these projects were not very successful. Before the end of 1905, Eddington had made the move to astronomy and in this field he made enduring contributions. Astronomy

[c]Information on Eddington's life is abridged from an article by J. J. O'Connor and E. F. Robertson.

Fig. 7.8 Sir Arthur Eddington (1882–1944), the leading astrophysicist of his day. He was the first to realize that the Sun's energy derived from thermonuclear fusion and predicted that it would live for 15 billion years. It is now into its life by 4.5 billion years.

had interested him from childhood ever since he was given access to a 3-inch telescope.

When only 31, Eddington was appointed to fill the chair of Plumian Professor of Astronomy left vacant upon the death of George Darwin, a son of Charles Darwin. In this new post he became director of the Cambridge Observatory, taking on the responsibility for both theoretical and experimental astronomy at Cambridge. Soon afterward he was elected a Fellow of the Royal Society.

Shortly after taking up his post at Cambridge, World War I broke out. Coming from a Quaker tradition and, as a conscientious objector, he avoided active war service and was able to continue his research at Cambridge during the war years of 1914–18. In 1920, Arthur Eddington was the first to suggest that stars obtained their energy from nuclear fusion of hydrogen to helium.

Eddington (Figure 7.8) became interested in General Relativity in 1915 when he received papers by Einstein and by de Sitter, which came to him through the Royal Astronomical Society. His particular interest was aroused because it provided an explanation for the previously noticed advance of the perihelion of Mercury. General Relativity predicted twice the amount given by Newton's theory of gravity. Einstein, himself had made this calculation as a test of his theory.

At 36, Eddington led an expedition (1919) to Principe Island in West Africa to observe the solar eclipse and to measure the degree to which light passing close to the Sun was bent. He sailed from England in March 1919 and by mid-May had his instruments set up on Principe Island where full eclipse was due at two o'clock in the afternoon of 29 May. That morning there was a storm with heavy rain and the clouds interfered with the photography of the eclipse. Nevertheless, Eddington wrote of the last few photographs "I hope [they] will give us what we need..." The results from Eddington's African expedition provided the first confirmation of Einstein's theory that gravity will bend the path of light when it passes near a massive star by *twice* as much as Newtonian physics predicts; the story made headline news in London and New York. Whereas Newtonian theory accounts only for the effect of the star's gravity on light, Einstein's theory predicts an equal and additional bending because of the *curvature* of spacetime.

Eddington lectured on relativity at Cambridge, giving a beautiful mathematical treatment of the topic. He used these lectures as a basis for his book *Mathematical Theory of Relativity* which was published in 1923. Einstein said that this work was "the finest presentation of the subject in any language."

Eddington was knighted in 1930 and received the Order of Merit in 1938. He received many other honors including gold medals from the Astronomical Society of the Pacific (1923), the Royal Astronomical Society (1924), The National Academy of Washington (1924), the French Astronomical Society (1928), and the Royal Society (1928). In addition to election to the Royal Society, he was elected to the Royal Society of Edinburgh, the Royal Irish Academy, the National Academy of Sciences, the Russian Academy of Sciences, the Prussian Academy of Sciences and many others.

7.5. Eventual Fate of Our Sun

Sir Arthur Eddington, correctly guessed in the 1920s that a star shines because, under the enormous pressure of its own weight, the core is heated to such high temperatures that hydrogen is burned to form helium.[d] He realized that our Sun will shine as we see it now for a total of 12 billion years before consuming all its primordial hydrogen; the thermal and radiant pressure produced by the fire will resist the attraction of gravity and forestall the ultimate collapse of the star.

[d]The process is known as thermonuclear burning; the mass-energy of helium is less than that of the four hydrogen and the excess appears as heat and radiation.

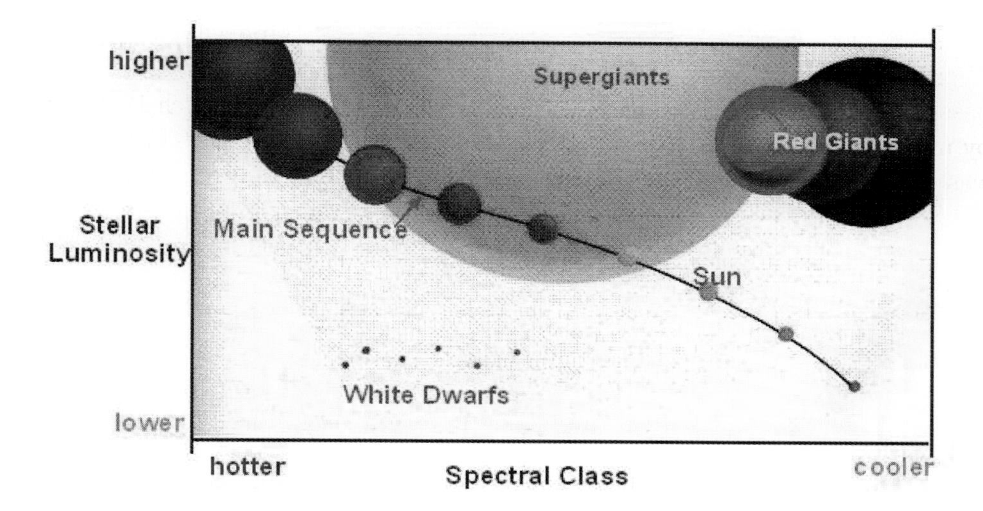

Fig. 7.9 The brightness of stars shown on the left scale is compared with their temperature which, for stars on the main sequence (marked) increases with mass shown on the bottom scale. Late stages in the evolution are shown by the giant stars at the top of the figure. Final stages of white dwarf (and neutron star) lie near the bottom. The most massive stars (top) make excursions near the end of their lives as they expand through the red giant stages, eventually dying in a supernova explosion. The Sun also will puff up through several stages but its red giant stage will more quietly subside into a white dwarf surrounded by an expanding planetary nebula. The relations shown here are referred to as a Hertzsprung-Russell diagram after its discoverers. Credit: UC Berkeley.

Eventually however, the Sun (Figure 7.9) will go through several stages lasting for several hundred million years during which it will burn hydrogen to form helium in the *outer* layers and expand enormously because of the heat released by these reactions; it will temporarily form a red giant, which will engulf the Earth and inner planets. Meanwhile the core will continue to heat during its contraction by conversion of gravitational energy to heat energy; the thermonuclear burning continues beyond helium to carbon with a release of energy that reheats the outer layers as the star expands as a *Red supergiant*, extending beyond Mars. At this point explosions in the core will cast off most of the star and the remaining small core will subside to form a white dwarf surrounded by an expanding shell of hot gases, in what is called a *planetary nebula*, as in Figures 7.10 and 7.11.

7.6. The Planets

Our solar system is about 4.5 billion years old, having formed about two-thirds of the way into the life of the universe. Eight *major* planets were

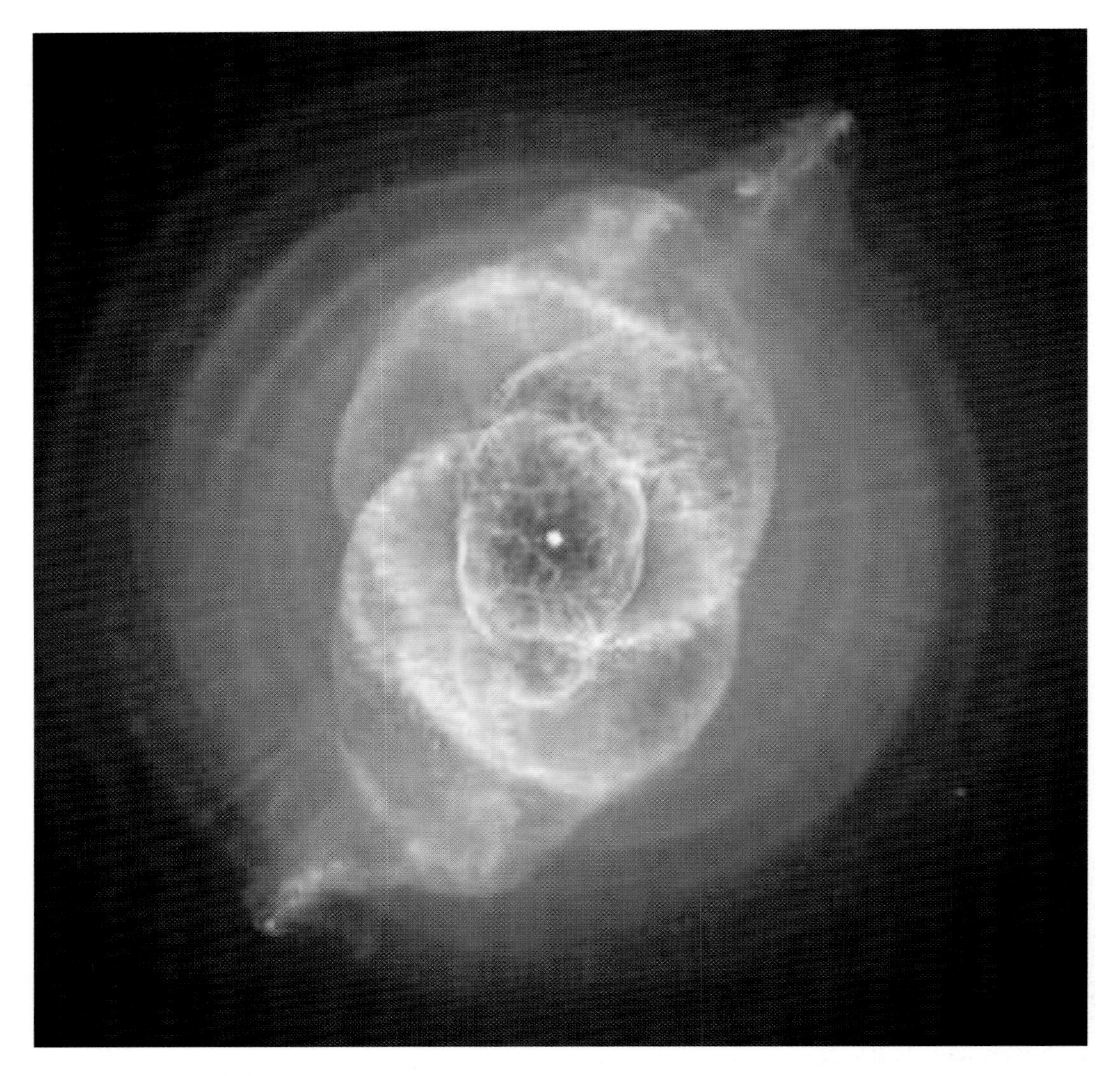

Fig. 7.10 A view of the Cat's Eye Nebula (NGC6543). A small dying star of the approximate mass of our Sun is thought to inhabit the center of the nebula, having cast off the gases from the central star over a succession of long intervals of an estimated 1,500 years between outbursts. It is the most complex of *Planetary Nebula*. Credit: NASA, ESA, HEIC, and the Hubble Heritage Team (STScI/AURA). Acknowledgments: R. Corradi (Isaac Newton Group of Telescopes, Spain) and Z. Tsvetanov (NASA).

formed. Four of them, the inner ones, Mercury, Venus, Earth and Mars, have roughly the same density and are made mostly of rock and iron and are known as the *rocky planets* or as the *terrestrial planets*. Between these planets and the outer ones lies an asteroid belt. Asteroids range in size

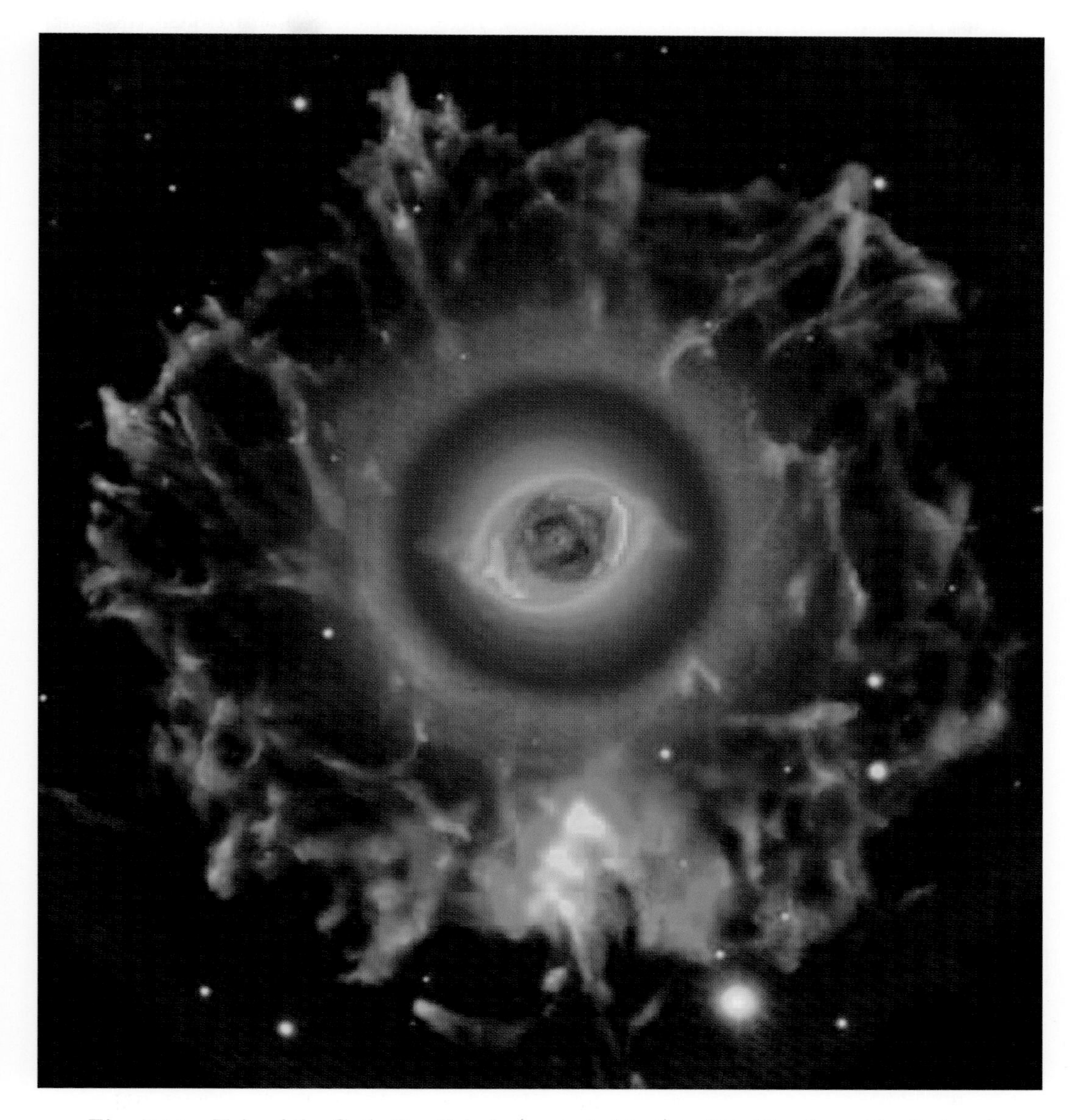

Fig. 7.11 Halo of the Cat's Eye Nebula (central object) is 27 trillion km wide. It dwarfs the Cat's eye nebula itself, extending far beyond that of Figure 7.10, which in this view occupies the small region at the center. Made with data from the Nordic Optical Telescope in the Canary Islands, the composite picture shows emission from nitrogen atoms as red and oxygen atoms as green and blue shades. Credit: R. Corradi (Isaac Newton Group) and D. Gonçalves (Inst. Astrofisica de Canarias).

from 1000 km (625 miles) across, to mere dust particles, but in mass they comprise less than 1/1000 of Earth's mass. Saturn and Jupiter are giant gas planets with masses of 100 and 300 times the mass of Earth, respectively. Uranus and Neptune are about 15 Earth masses made mostly of water, while the far-off miniscule *dwarf planet* Pluto, with a mass of 1/500 that of Earth, is made of ice and rock. It is now known as a dwarf planet — not one of the (now) eight *major* planets of the solar system.

Most — but not all — planets have moons. Altogether, 139 moons have been discovered in our solar system so far. While Earth has only one, Jupiter has at least 60 moons. Four of them were discovered by Galileo shortly after he built his own telescope; the latest ones in our own day by astronauts. Galileo named the four moons that he had discovered, the Medician moons, hoping to remain in the favor of the House of Medici, which ruled Florence for more that 300 years. The family's founder, Cosimo Medici, a patron of the arts and ruler of Florence except in name, died 100 years before Galileo was born. Although Cosimo shunned the public light, his descendants became dukes and duchesses, the rulers of the state of Tuscany.

7.7. The Earth

The Earth and other planets formed in the disk of the Milky Way Galaxy. The galaxy is a thin sheet of gases composed mostly of hydrogen and helium with a sprinkling of heavy elements together with dust and molecules including water that is interspersed among the stars. The Milky Way Galaxy has the shape of a disk with a central bulge and is about 100,000 light-years across.[e] The solar system — our Sun and planets — lie in the disk about 2/3 distant from the center (Figure 7.2).

The Sun formed first, and then attracted dust and molecules which collided with each other to form larger bodies called *planetesimals* ranging in size from 1/10 to 10 km. These were massive enough to gravitationally attract each other and they grew, eventually to form the planets.[f]

The interior of the Earth was heated and partially melted by the impact energy of the planetesimals from which it formed, and also by short-lived radioisotopes, mostly a neutron rich form of aluminum that was produced by supernova near the Solar System at the time of its birth. Long-lived

[e] A light-year is the distance light travels in a year and is approximately 10^{18} km = billion-billion km.

[f] A 1/10 km radius planetesimal would weigh about 40,000 billion pounds assuming it to have the average density of Earth.

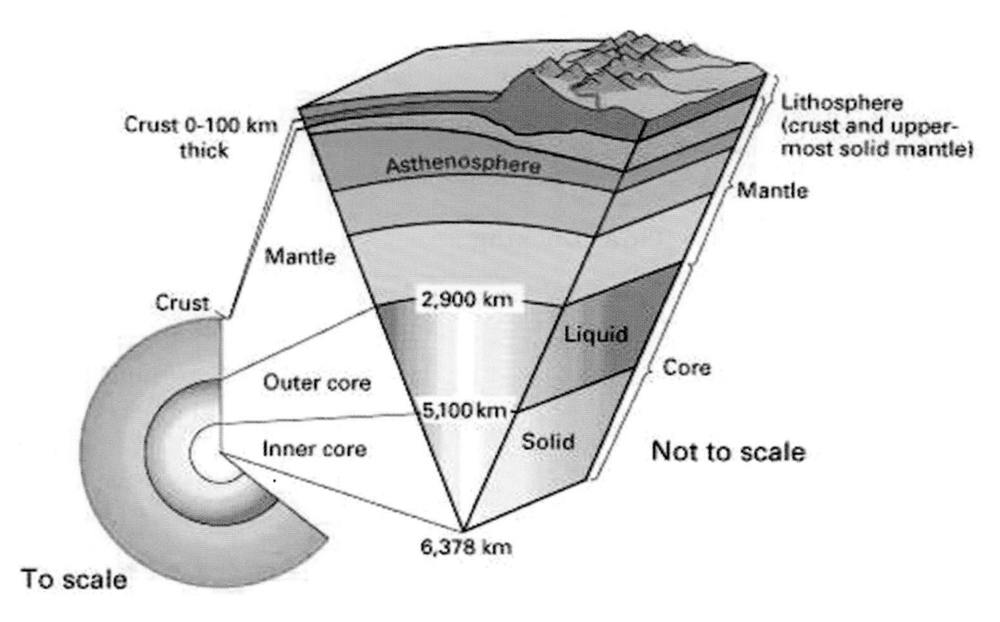

Fig. 7.12 The Earth's interior showing hot interior, the core solidifies by pressure of the outer Earth, surrounded by liquid which is in convective motion as it carries heat from the inner core outward. Notice the weight of mountains thickens the crustal regions and the thinner crust under the lighter weight of water. Credit: US Geological Survey and John M. Watson.

radioactivity has sustained high temperatures in the interior. Out to about half the Earth's radius, temperatures are so high that under normal circumstances this region would be liquid; however, a small central region is maintained in a solid state by the high pressure exerted by the weight of all that lies above. Outside this region a liquid core extends to about half the Earth's radius. Above that sits a solid mantle and at the very top, a thin crust comprising plains, mountains, and ocean floor. The crust is thin under the oceans and thick under mountains (Figure 7.12).

Although the central core is above normal melting temperature, the weight of the overlaying layers solidifies it. The transport of heat by the flow of the molten fluid above the solid core to the mantle and its resettling as it cools, called convection, together with the rotation of the Earth, is probably responsible for Earth's magnetic field. Venus with a similar iron-core composition but a very slow rotation compared to Earth (one revolution every 243 Earth-days) does not have a measurable magnetic field. The field of the Earth deflects electrons flowing from the Sun's surface to the poles, where it creates the aurora borealis (called Northern Lights by those of us in the northern hemisphere). This is more than a beautiful display of lights; it is the visible manifestation of the protective role played by the Earth's atmosphere and magnetic field in absorbing and deflecting harmful radiation from the Sun.

7.8. The Moon and Tidal Locking

The Moon's face drawn by Galileo in 1609 and shown on page 157 reveals the same face we see today. And most likely it is the same face that the moon turned to Earth when early man gazed at the night sky, and for millions of years before him. Why does the Moon always present the same face to us on Earth? This is caused by the effect of gravity acting over a long period of time. It was not always so. Long ago the Moon must surely have been rotating, as all heavenly bodies seem to. Why does it now always face the Earth?

The near side of the Moon is attracted by the Earth's gravity more strongly than the far side whether or not it is rotating. As a result, the Moon's shape is distorted from that of a perfect sphere; it is elongated in a direction along the axis joining it to the Earth, a shape that is called an ellipsoid.

Meanwhile, the Moon was rotating so that its interior matter was constantly forced to rearrange itself so as to retain the orientation of the ellipsoidal shape pointing always toward Earth. To rearrange the interior, the viscosity and friction of the interior matter — at first liquid, and later rocky — had to be overcome. Over millions of years this continuous rearrangement sapped energy from its cause — rotation. Some rotational energy was thereby converted into heat and possibly caused melting of the interior in early times. As a consequence of this tidal effect, the Moon in its orbit about Earth ceased to rotate freely about its own axis, but rather always shows the same face to Earth. The Moon still does rotate but now very slowly, once a lunar month, in synchrony with its orbit about Earth. This phenomenon is often referred to as *tidal locking*. Tidal locking is also observed in some of the moons of Jupiter, and in particular the interior of Io is molten, made so by the tides induced by Jupiter. The molten interior gives rise to intense volcanism on that moon.

For the same reason, the Earth is slightly deformed by the Sun's gravity, and eventually may present only one face to the Sun, leaving one side in perpetual day, and the other in perpetual night. Depending on how long this takes, the whole Earth and inner planets may have been already engulfed

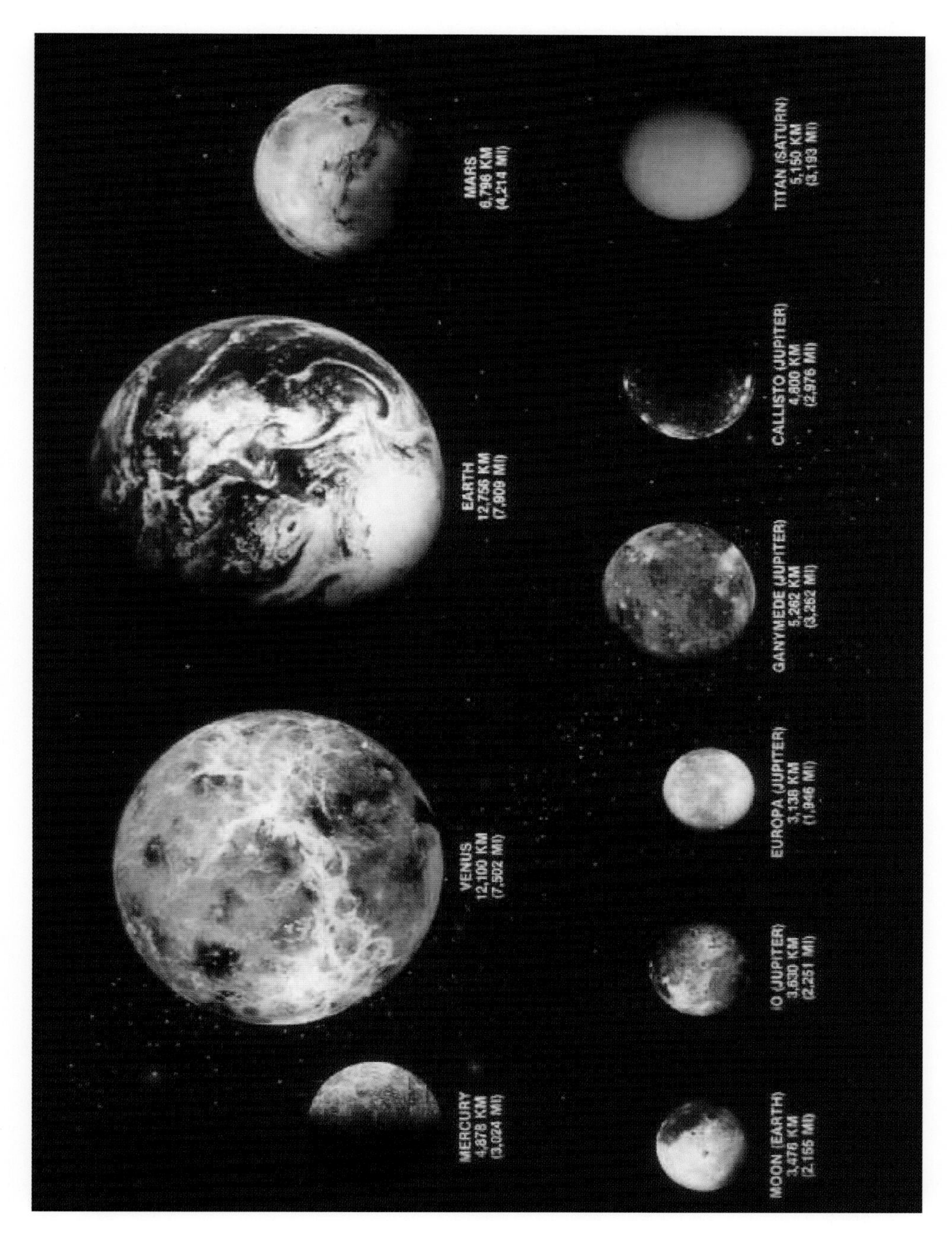

Fig. 7.14 Comparison of planet sizes (top) and some of their moons (bottom). Credit: NASA.

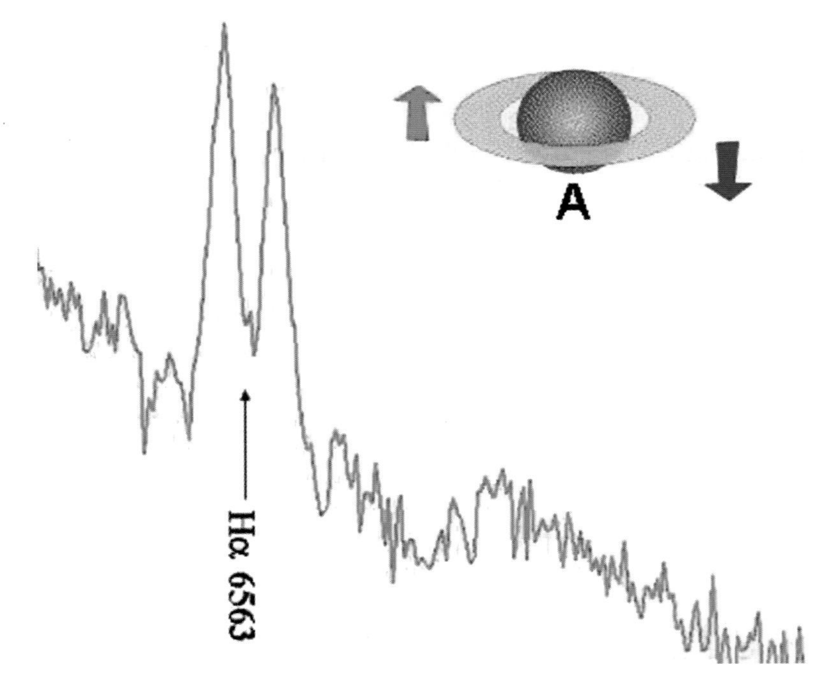

Fig. 7.15 One spectral line of light emitted by hydrogen is spread into two, corresponding to light received from the two edges of a rotating star. The retreating edge emits a red-shifted line and the advancing edge, a blue-shifted line. This is analogous to the Doppler shift of a retreating and approaching train whistle. Credit: Osservatorio Astronomica "G.V. Schiaparelli".

7.12. Box 19

19 Technical Note on Dating the Formation of Solar System

The era of formation of the solar system from material in the disk of the early Milky Way Galaxy, can be inferred from the presence of ^{26}Mg, a product of the radioactive decay of the long-lived excited state of the isotope of ^{26}Al (half life of $\approx$ 0.73 million years) in chondritic meteorites. Chondrite is a type of stony meteorite made mostly of iron and magnesium-bearing silicate minerals that have remained little changed from the dawn of the Solar System, about 4.5 billion years ago.

Chondrites are the commonest kind of meteorite, accounting for about 86% of them, and, aside from the primordial elements, hydrogen and helium, they have the same elemental composition as the original solar nebula because they come from asteroids that had never melted or underwent differentiation.

The occurrence of ^{26}Al decay products in the inclusions shows that the clouds from which the solar system condensed, formed while that isotope was still present, so that agglomeration took place over a very short timescale, less than 3 million years. The most abundant component within chondrites, viz. chondrules, show little evidence for the presence of radioactive ^{26}Al, implying that chondrule-forming process took place about 2–3 million years or so after the formation of Ca-Al-rich inclusions.

The ^{41}Ca–^{41}K chronometer, with half life $T_{1/2} \approx 0.15$ million years, implies even more rapid formation of collapsing interstellar clouds, with an interval between nucleosynthesis and agglomeration of less than 0.3 million years. Absolute dating of the components within meteorites uses the U-Pb isotope system, and places the date of formation of these clouds at 4.566 billion years, the estimated life of the Sun.

Chapter 8

First Modern Astronomer

In questions of science, the authority of a thousand is not worth the humble reasoning of a single individual.

Quoted in Arago, *Eulogy of Galileo* (1874)

8.1. Galileo: Principle Discoveries

Galileo Galilei (1564–1642) is well-known for his trial before the inquisition in Rome, but perhaps less well for his writings about the marvelous discoveries he made with the telescope, which he greatly improved, though did not invent. He is even less well-known for his work on the microscope, and his investigations in ornithology and botany to which he contributed as a member of the Lincean Academy in Rome together with his astronomical studies.

The Lincean Academy (*Accademia dei Lincei*), one of the first learned societies, had been founded by the young Prince Federico Cesi (1585–1630) at the age of 18. This young man who died at 45, had he lived to the age of Galileo, would likely have become to natural history, what Galileo became to the physical sciences. In fact, Galileo himself was inducted into the membership of the Lincean Academy in 1611; he was the sixth member. Galileo was very proud of his membership and even had this fact noted, among his other posts and distinctions, on the title page of several of his books. Cesi was a staunch supporter of Galileo but died before he could be of help through his later troubles in Rome.

Galileo was probably the first to turn a telescope to the heavens (1609), having already made many observations by sight. He soon learned by observation that the starry heavens were not fixed, for he discovered four of the moons of Jupiter and noted that their positions and spacing changed; sometimes they disappeared one by one behind (or in front of) Jupiter, reappearing later on the other side. After recording the motion of these

Fig. 8.1 Galileo Galilei, first astronomer to turn a telescope to the heavens; thus began a revolution in understanding our place in the universe. Image credit and permission: Instituto e Museo di Storia della Scienza, Firenze, Italia.

moons over a period of time on many occasions he deduced that they orbited *Jupiter*. This was counter to Church doctrine, which was fixed by the ideas of Aristotle on such matters — Earth was said to be the center of all moving bodies. Galileo named the heavenly objects that he had discovered, the "Medicean moons" after his patrons, the heirs of the Florentine banker Cosimo Medici, il Vecchio (1389–1464), father of the most powerful patrician family in Florence, patron of the artists Ghiberti, Brunelleschi, Donatello, Alberti, Fra Angelico, and grandfather of Pope Clement VII.

Galileo was further persuaded that the Earth was not the center of the universe by his observation of the phases of Venus, being sometimes fully illuminated, then gradually fading on one side, becoming a mere sliver, and then lighting on the other side until it was fully illuminated again. From this he deduced that Venus was a planet that circled the Sun — not the Earth. The phases of the Moon he adduced as further evidence, and noted that its surface was not the *perfect celestial sphere* attributed to it by Aristotle, whose beliefs the church had adopted; rather, it had mountains and craters that were revealed by the shadows cast by the Sun (Figures 8.2 and 8.3). These discoveries he wrote about in 1610 for the educated public in a marvelous small book, *Sidereus Nuncius* (Starry Messenger), which caused a sensation upon its appearance. To this day it is informative and well worth reading (see Bibliography).

8.2. Youth

Galileo Galilei was born in February 1564 into a gifted musical family whose home was in the countryside near Pisa (Italy). He died in January at the age of 78 at his hillside villa in Ercetri, above Florence, and overlooking the Arno River. In his last year, officials of the Inquisition kept him under house arrest, even preventing him at times from visiting his doctor in Florence. Galileo died in the same year that Isaac Newton was born. Galileo Galilei's parents were Vincenzo Galilei and Guilia Ammannati. Vincenzo, who was born in Florence in 1520, was a music teacher and lutist.

When Galileo was eight years old, his family returned to Florence but Galileo remained with a relative of his mother in Pisa for another two years. At the age of ten, Galileo left Pisa to join his family in Florence. Once he was old enough, he was sent for his early education to the Camaldolese Monastery on a magnificent forested hillside at Vallombrosa near Florence. The Order combined the solitary life of the hermit with the strict life of the monk and soon the young Galileo found this life an attractive one. He became a novice, intending to join the Order. However, his father Vicenzo had already decided that his eldest son should become a medical doctor and he was sent back to Pisa for a medical degree. But Galileo also attended courses on mathematics and natural philosophy — attending a course on *Euclid's Elements* taught by Ostilio Ricci, mathematician of the Tuscan Court — the descendants of Cosimo Medici. During the summer Galileo

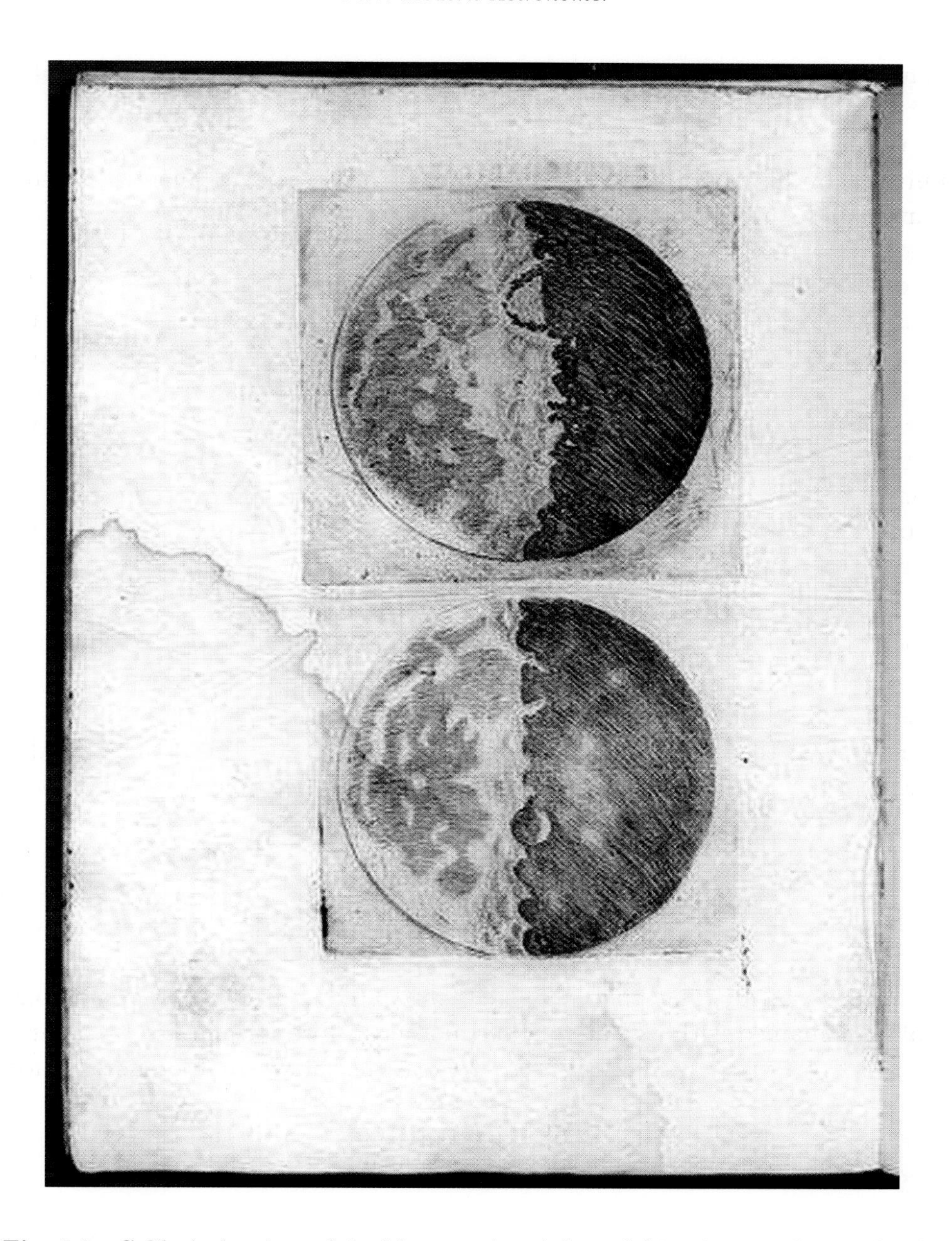

Fig. 8.2 Galileo's drawings of the Moon as viewed through his telescope showing by the shadows cast by the Sun at two different times how the Moon's surface exhibits mountains and valleys — not the smooth *perfect sphere* envisioned by Aristotle and adopted as Roman Church belief. From an original copy of his 1610 book *Siderius Nuncius* (Starry Messenger). Credit: Exhibit from the College of Arts and Science, University of Oklahoma; Kerry Magruder, with the assistance of, Marilyn B. Ogilvie, Duane H. D. Roller. Image permission: ©History of Science Collections, University of Oklahoma Libraries.

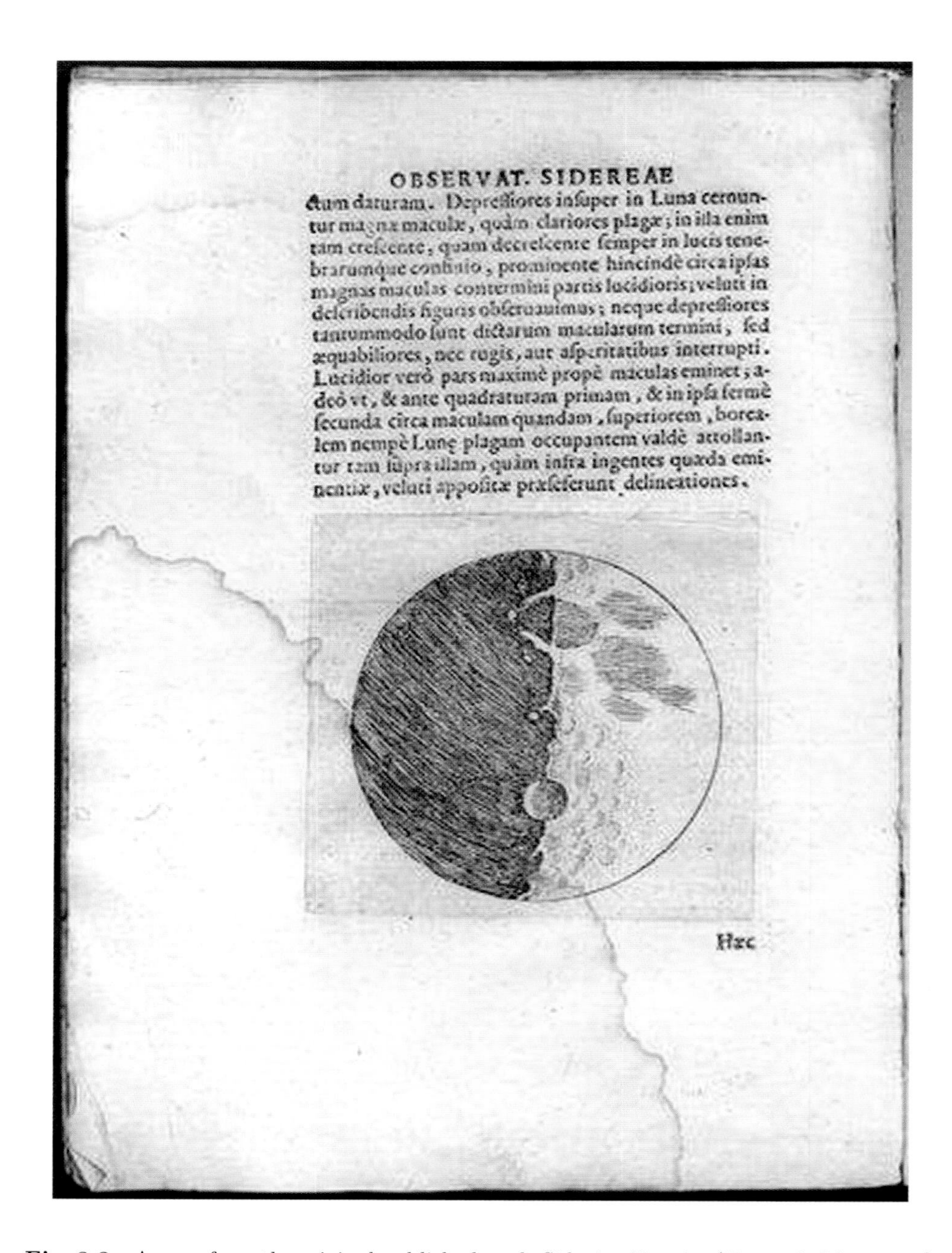

OBSERVAT. SIDEREAE

ctum daturam. Depreſſiores inſuper in Luna cernuntur magnæ maculæ, quàm clariores plagæ; in illa enim tam creſcente, quam decreſcente ſemper in lucis tenebrarumque confinio, prominente hincindè circa ipſas magnas maculas contermini partis lucidioris; veluti in deſcribendis figuris obſeruauimus; neque depreſſiores tantummodo ſunt dictarum macularum termini, ſed æquabiliores, nec rugis, aut aſperitatibus interrupti. Lucidior verò pars maximè propè maculas eminet; adeò vt, & ante quadraturam primam, & in ipſa fermè ſecunda circa maculam quandam, ſuperiorem, borealem nempè Lunę plagam occupantem valdè attollantur tam ſupra illam, quàm infra ingentes quædam eminentiæ, veluti appoſitæ præſeferunt delineationes.

Hæc

Fig. 8.3 A page from the original published work *Siderius Nuncius* (Heavenly Messenger) by Galileo (1610). By comparing the shadows cast by the Sun at different phases as in Figure 8.2, one can see the evidence that persuaded Galileo that there are deep valleys on the surface of the Moon. Permission: ©History of Science Collections, University of Oklahoma Libraries.

was back in Florence with his family and invited Ricci to meet his father. Through Ricci's persuasion, Galileo was allowed to study the works of Euclid and Archimedes from the Italian translations.

Galileo began teaching mathematics, first privately in Florence and then at Siena. At the age of only 22 he taught at Vallombrosa, and in this year he wrote his first scientific book. In the following year he made the long journey to Rome to visit Clavius, professor of mathematics at the Jesuit *Collegio Romano* there. This was the first of six long visits to Rome, the first as a student, the second as the renowned author of the book recounting his discoveries with the telescope, feted by the nobles and Cardinals of the Church, housed in luxurious villas, and given six audiences with the pope. His last visit as an old man was under a cloud cast by the Church because of his belief in the Copernician system, which held that the Earth and other planets moved about the Sun.

At the time of his first visit, centers of gravity of solid bodies was a popular topic and Galileo brought with him results which he had discovered. He maintained a correspondence with Clavius and the theorems that Galileo had proven on the centers of gravity were discussed in their correspondence. It is also likely that Galileo received lecture notes from courses which had been given at the *Collegio Romano*, for he made copies[a] of such material which survive to this day. At the age of 24, Galileo received a prestigious invitation to lecture at the Academy in Florence on the dimensions and *location of hell* written of in Dante's Inferno. This must have required some (perhaps tongue in cheek) ingenuity for the man who was father of the scientific method.

The University of Pisa appointed Galileo to professor of mathematics at the age of 25. His father died two years later. Because Galileo was the eldest son, he became responsible for supporting the rest of the family and providing dowries for his two younger sisters. To this end, Galileo sought a more lucrative post and at the age of 28 he was appointed professor of mathematics at the University of Padua at a salary of three times what he had received at Pisa.

He spoke of these eighteen years at Padua as the happiest of his life. Not only did he teach Euclid's geometry and standard (geocentric) astronomy to medical students, who would need to know some astronomy in order to make use of astrology in their medical practice: He also carried on his own intellectual pursuits. Galileo argued against Aristotle's view of astronomy and natural philosophy in three public lectures he gave in connection with

[a]No Xerox then; copies were by hand.

the appearance in 1604 of a New Star (now known as *Kepler's supernova*, Figure 8.4). A reproduction from one of Kepler's books locating this supernova (SN 1604) is shown in Figure 8.5. The belief of Aristotle, which had been adopted as doctrine by the Roman church, was that all changes in the heavens had to occur in the lunar region close to the Earth; the realm of the fixed stars was an everlasting enveloping canopy. But Galileo used parallax arguments to prove that the New Star could not be close to the Earth. In a personal letter written to Kepler in 1598, Galileo stated that he believed the theories of Copernicus who discovered that the planets circled the Sun — not the Earth. However, no public sign of this belief was to appear until many years later.

At Padua, Galileo began a long-term relationship with Maria Gamba, who was from Venice, but they did not marry perhaps because Galileo felt his financial situation was not secure enough. In 1600 their first child Virginia was born, followed by a second daughter Livia in the following year. In 1606 their son, Vincenzo, was born.

8.3. His Astronomical Research with the Telescope

Galileo was 45 when he received a letter from Paolo Sarpi, who had been informed by way of diplomatic pouches, telling him about a spyglass made by a Dutchman, Hans Lippershey. Galileo made his own telescope soon after its invention in 1608. The first ones made in Holland and Paris used lenses made by opticians for eye-glasses and had a magnification of only three. Galileo was determined to make his own more powerful instrument. After some experimentation, he learned that he needed lens which were not available in optical shops. He therefore set about teaching himself the art of grinding lenses. Soon Galileo turned his own more powerful telescopes toward the heavens, realizing that this new instrument would revolutionize astronomy and cosmology.

After perfecting an instrument that would magnify twenty times he began numerous systematic observations which he recorded in drawings, notes, and published books.[b] One of them, *Discourses on two New Sciences*, was written in 1638 after his trial before the Inquisition. The manuscript was smuggled out of Italy to Leiden for publication.

[b]Galileo Galilei (1564–1642): *Sidereus Nuncius*, published in Venice, May 1610; translated by Albert Van Helden, The University of Chicago Press, 1989: *Dialogue Concerning the Two Chief World Systems*, published in Florence, February 1632, translated by S. Drake, 2nd edition 1967, University of California Press: *Discourses on two New Sciences*, published in Leiden, 1638; translated by S. Drake, 1974, The University of Wisconsin Press.

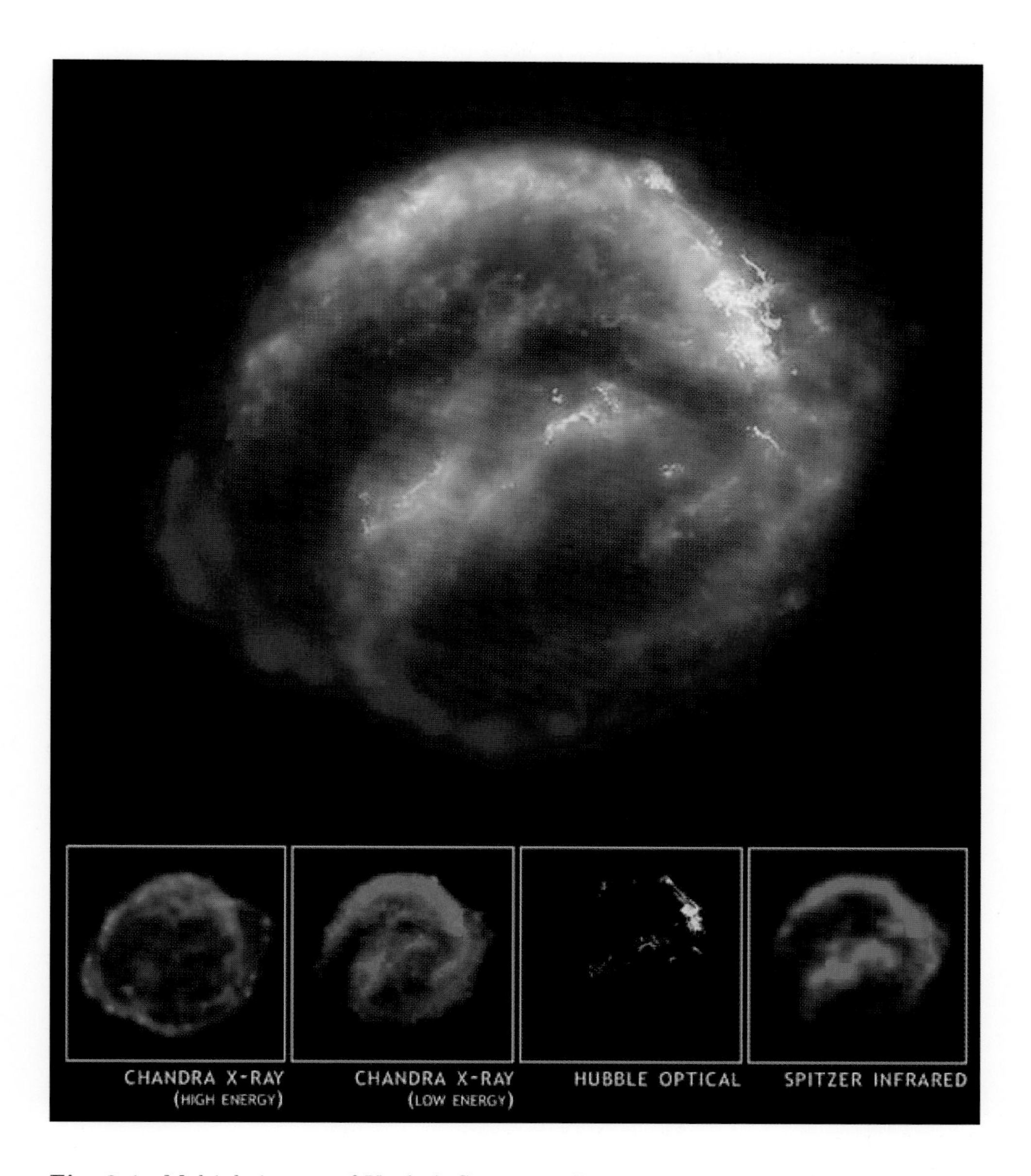

Fig. 8.4 Multiple images of Kepler's Supernova Remnant; the Hubble Space Telescope, the Spitzer Space Telescope, and the Chandra X-ray Observatory — joined forces to probe the expanding remains of a supernova. Now known as Kepler's supernova remnant, this object was first seen 400 years ago by sky watchers, including the famous astronomer Johannes Kepler. The combined image unveils a bubble-shaped shroud of gas and dust that is 14 light-years wide and is expanding at 4 million miles per hour (2,000 km per second). Credit: NASA/ESA/JHU/R. Sankrit and W. Blair.

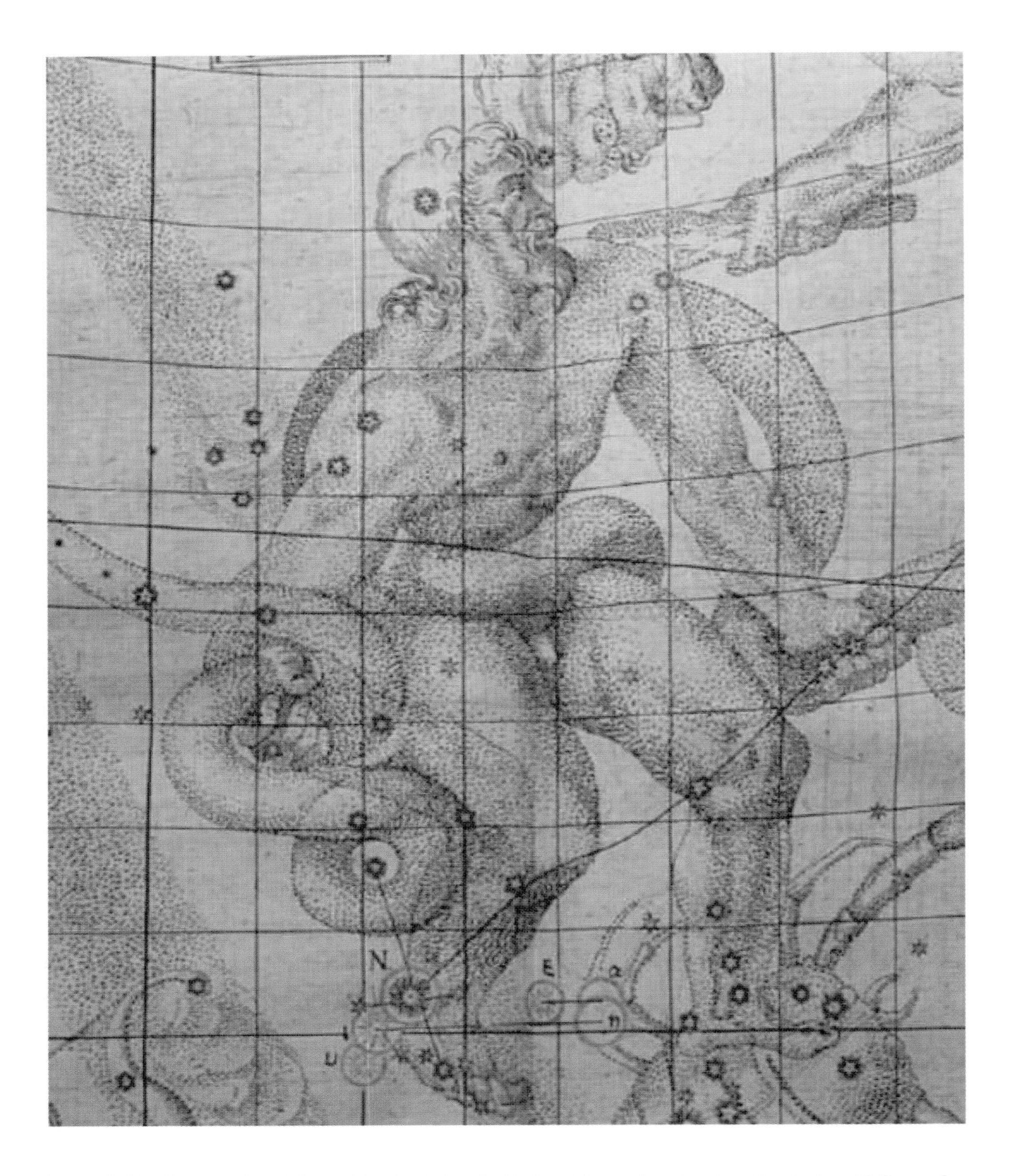

Fig. 8.5 A star chart from Kepler's book *Stella Nova* showing by the letter "N" at the foot of the man's figure the location of Kepler's supernova of 1604.

In about two months, December and January of 1609, Galileo made more discoveries that changed the worldview than anyone before him. Indeed, he opened the universe as a realm for study, which we continue to this day with

LIGO / Caltech

Fig. 8.6 An aerial photograph provides a sense of the massive scale of the Laser Interferometer Gravitational-Wave Observatory on the Hanford Nuclear Reservation. Eventually, the L-shaped facility will stretch 2.5 miles (4 km) in each direction.

ever more powerful and varied telescopes, viewing the universe to its deep reaches over a wide range of radiation, from x-rays through to optical and on to the infrared, as for example in Figure 8.4. Some of the instruments are carried aloft in spacecraft. Soon, astronomers hope to detect very distant large-scale disturbance in the universe that make waves on the very fabric of spacetime (Figure 8.6). They are called gravitational waves and their possible existence was predicted by Einstein, though he doubted that they would ever be measurable.

Galileo had also sent Cosimo de Medici,[c] the Grand Duke of Tuscany, an excellent telescope for himself. He succeeded in impressing Cosimo and, in June 1610, only a month after his famous book was published, became Chief Mathematician at the University of Pisa (without any teaching duties) and 'Mathematician and Philosopher' to the Grand Duke of Tuscany. In 1611 at the age of 47 he visited Rome where he was treated as a leading celebrity; the *Collegio Romano* put on a grand dinner with speeches to honor Galileo's remarkable discoveries. He was also made a member of the *Accademia dei Lincei* and this was an honor that was especially valued by Galileo.

[c]An heir of Cosimo il Vecchio, banker and great patron of Florence and the arts.

8.4. The Earth Turns Round

While in Rome, and after his return to Florence, Galileo continued to make observations with his telescope. Already in the *Starry Messenger* he had given rough periods of the four moons of Jupiter. But more precise calculations were certainly not easy; it was difficult to identify from only several observations, which Moon was which, because of their eclipse several at a time. After a long series of observations he was able to give accurate periods by 1612. At one stage in the calculations he became very puzzled since the data he had recorded seemed inconsistent, but he had forgotten to take into account the motion of the Earth round the Sun, a mistake that has been repeated in a preliminary report concerning exoplanet discoveries even in modern times.

Galileo knew that all his discoveries were evidence in favor of the observations and views put forward by Copernicus. Cardinal Robert Bellarmine was the Church's most important figure at this time concerning interpretations of the Holy Scripture. He saw little reason for the Church to be concerned regarding the Copernican theory. The point at issue was whether Copernicus had simply put forward a mathematical theory that enabled the calculation of the positions of the heavenly bodies to be made more simply or whether he was proposing a physical reality. Bellarmine viewed the theory as an elegant mathematical one, which did not threaten the established Christian belief regarding the structure of the universe.

Galileo in his maturity had many powerful friends, even several successive Popes. Among his friends was the Grand Duchess Christina of Lorraine, a descendent of Cosimo Medici (il Vecchio), with whom he frequently corresponded on matters concerning his research. In these letters he argued strongly for a non-literal interpretation of Holy Scripture when the literal interpretation would contradict facts about the physical world proved by observation. Men of religion and those of science have often been at odds, not having the wisdom of Galileo to realize that science concerns the physical world, and religion, the spiritual world.

To the Duchess, Galileo also wrote quite clearly that for him the Copernican theory about the motion of the planets around the Sun rather than the Earth was not just a mathematical calculating tool, but a physical reality. This became a focus of charges made against him at the inquisition years later.

Copernicus's writings concerning his celestial observations were gaining widespread notice throughout Europe. This became increasingly alarming to the hierarchy of the Roman church. Pope Paul V ordered Cardinal Bellarmine to have the Sacred Congregation of the Index decide on the Copernican theory. The cardinals of the Inquisition met in 1616 and took evidence from *theological* experts. They condemned the teachings of Copernicus, and Bellarmine conveyed their decision to Galileo. Galileo was forbidden to hold Copernican views but later events made him less concerned about this decision of the Inquisition. Most importantly Maffeo Barberini, who was an admirer of Galileo, was elected as Pope Urban VIII. This happened just as Galileo's book *Il saggiatore* (The Assayer) was about to be published by the Lincean Academy in 1623. Thereupon, Galileo dedicated this work to the new Pope. It described Galileo's new scientific method.

Pope Urban VIII invited Galileo to papal audiences on six occasions at which he was friendly and showed much interest in Galileo's ideas. Galileo left with the belief that the Catholic Church would not make an issue of the Copernican theory except that a few lines should be deleted from his manuscript. Galileo, therefore, decided to publish his views believing that he could do so without serious consequences from the Church. Indeed, the records from that time are very unclear.[d] But by this stage in his life Galileo's health was poor. He had frequent bouts of severe illness and so even though he began to write his famous Dialogue in 1624 it took him six years to complete the *Dialogue concerning the two chief world systems.* It is written as a discussion between three men concerning the Copernican world view and that of the ancient one of Ptolemy. Devastating remarks were made in it, also, of the Aristotelian view, the view held by the Roman Church in matters of cosmology. Galileo's book was banned by the Church and remained on the list for 200 years.

8.5. Conflict with the Church

Galileo attempted to obtain a license from Rome to publish the Dialogue. However, the head of the Lincean Academy, Federico Cesi — who would normally have expedited approval of a license and taken care of the cost of printing as he had for other works of Galileo — died. The Church, through its many functionaries and reviewers kept very tight control of what could be printed. Galileo would have had to send the manuscript to Rome for review. This he was apparently willing to do, but was advised by the Tuscan Secretary of State that it could not be done safely because of the conditions

[d]Cf. the works by Finocchiaro and by Sharratt in the Bibliography.

of the roads, the toll gathers, and thieves. This was a very important consideration: all manuscripts at that time were written by hand, and making copies was a laborious task.

Such quibbles, as they may seem to us, as "one would never be admitting the *absolute* truth of this opinion, but only its *hypothetical* truth without benefit of scripture [emphasis mine]" were a great thorn in Galileo's side.

Eventually he received permission to publish from officials of the church in Florence. This left him open later to an accusation by the inquisitors of bypassing Rome. In February 1632 Galileo published the *Dialogue Concerning the Two Chief Systems of the World — Ptolemaic and Copernican.* Shortly after publication of the Dialogue, the Inquisition banned its sale and ordered Galileo to appear in Rome before them.

Illness prevented Galileo from traveling to Rome until 1633 when he was 69. Indeed, travel in those days was not easy under the best of circumstances. The Plague had broken out in several cities in 1630, including Florence and the road was dangerous. Galileo's accusation at the trial, which followed, was that he had breached the conditions laid down by the Inquisition in 1616. However a different version of this decision was produced at the trial rather than the one Galileo had been given at the time. The truth of the Copernican theory was not an issue therefore; it was taken as a fact at the trial that this theory was false. Indeed, the judgment of 1616 had declared the theory totally false.

Having been called to Rome, an arduous and dangerous journey for an old man, the proceedings against him took many months. He was sometimes threatened with imprisonment, or even of death if he did not obey the orders issued by officials of the Church. At various times he was under confinement in the villas of friends. Finally, Galileo was found guilty in 1633, made to abjure his belief in the Copernician system, and condemned to lifelong imprisonment.

> *I have been pronounced by the Holy Office to be vehemently suspected of heresy, that is to say, of having held and believed that the Sun is the center of the world.*
>
> Galileo, From his confession before the inquisition: sworn on the twenty-second day of June, 1633[e]

[e]Translated by Giorgio de Santillana, University of Chicago Press, Chicago, 1955, pp. 310–311.

However, the sentence was carried out more leniently and it amounted to house arrest rather than a prison sentence. He was permitted to live first with his friend, the Archbishop of Siena, of whom it is said, he saved Galileo's sanity, which was in peril because of his arduous trial and severe cross-examination before the Inquisition in Rome. Following his stay with the Archbishop, he was permitted to return to his home in Arcetri overlooking the Arno River near Florence. But Galileo had to spend the rest of his life watched over by unfriendly officers of the Inquisition. Even when sick, he was not always allowed to visit his physician in Florence. For such a great man, the only scientist who is recognized today by his *first name alone*, this was a sad and lonely end.

He suffered another severe blow when his daughter Virginia, Sister Maria Celeste, died. She had been a great support to her father through his illnesses and Galileo was so shattered that he could not work for many months. When he did restart work, he began to write *Discourses and mathematical demonstrations concerning the two new sciences.* After Galileo had completed work on the Discourses it was smuggled out of Italy, and taken to Leiden in Holland where it was published.

According to his will Galileo wished to be buried beside his ancestors in the family tomb in the Basilica of Santa Croce but his relatives feared that this would provoke opposition from the Church. His body was placed in a chapel of Sante Croce at that time, and only a hundred years later was it placed by the civil authorities with the approval and in the presence of the ecclesiastical authorities in a fine tomb in the Basilica Sante Croce. There, many visitors to Florence see it among other great figures of the Renaissance — Dante, Rossini, Michelangelo, and Machiavelli. Having ordered a thorough review of the papers regarding the Galileo affair, on 31 October 1992, 350 years after Galileo's death, Pope John Paul II gave an address on behalf of the Catholic Church in which he declared that errors had been made by the theological advisors in the case of Galileo. He declared the Galileo case closed.

Chapter 9

Life on Earth

If nature were not beautiful, it would not be worth knowing, and if nature were not worth knowing, life would not be worth living.

Henri Poincaré (1854–1912)

9.1. Land and Waters of Earth

No other planet in our solar system comes anywhere near the habitat that Earth provides. The essential ingredients for life exist on this planet, and literally every niche is occupied by some life form or other, from the depths of the oceans to the heights of the mountains. No other planet has an atmosphere that would support life, and some of the planets are extremely inhospitable, like the gaseous planets with poisonous atmospheres. Only four planets are terrestrial. And of these, only Earth has water and a thick atmosphere of oxygen and carbon dioxide. The thickness of the atmosphere protects life on Earth from harmful radiation from space. But it is neither so great that the pressure on the Earth's surface would overwhelm life as we know it, nor so insignificant that the Earth would lose heat almost as quickly as it is absorbed from the Sun. The atmosphere is thick enough that a significant greenhouse effect is produced; from the Sun, the Earth absorbs visible radiation which heats its surface; Earth's surface radiates some of this heat, raising the temperature of the carbon dioxide and water vapor in the atmosphere; a part of this heat is radiated back to Earth, the rest to space. On Earth the balance produces a hospitable mean temperature.

The oceans of the world cover almost three-quarters of its surface (Figure 9.1). Liquid water is a necessary ingredient of all forms of life on Earth. Yet of the land above the sea, much of it is under ice, and more still in desert such as almost all of North Africa (Figure 9.2). Nevertheless, life in some form or other exists from the deepest ocean troughs to the highest mountain tops. Aside from penguins, which for centuries have survived the

Fig. 9.1 This amazing photo of Earth taken by a European weather satellite, shows Africa, partly under cloud cover (white, pale mauve and turquoise wispy streaks). The northern part is desert. Much of Europe is under cover of clouds on this day. Oceans are deep maroon; vegetation shows up as green. At the far left, the coastline of south America can be viewed; at the right, the Red Sea and Saudi Arabia. Credit: ©2003 EUMETSAT (Europe's Meteorological Satellite Organization).

severe winter storms of Antarctica, primitive life even lives encapsulated in small cavities near the outer centimeter of rocky quartzites where sunlight can penetrate to promote photosynthesis and where water exists in its liquid state. Temperatures within such cavities can reach 10°C.

Water ice and water vapor can be found presently on some of the other terrestrial planets and satellites in our Solar System. And evidence for

Fig. 9.2 The Nile river in Egypt, near Luxor, showing the narrow strip of habitable and agricultural land beyond which stretches into the desert. Luxor (the ancient Thebes) is the location of the great necropolis known as the Valley of the Kings (circled). The Sun passed over the pharaoh's mortuary temple, built for their cult on the floodplain near Thebes, and then over the pharaoh's tomb in the valley, a symbolically important connection. Daytime photographs were taken from the International Space Station on 14 February 2003. Photograph courtesy of NASA-JSC Gateway to Astronaut Photography of Earth.

ancient seas and channels exist on Mars. As well, data from the Galileo mission suggest a layer of liquid water may exist beneath the icy surface of Europa, one of Jupiter's satellites. But Earth is the only planet in the solar system to have *liquid* water on its surface.

9.2. Conditions for Life on Earth

> *We have found a strange footprint on the shores of the unknown. We have devised profound theories, one after another, to account for its origins. At last, we have succeeded in reconstructing the creature that made the footprint. And lo! It is our own.*
>
> Sir Arthur Eddington (1920)

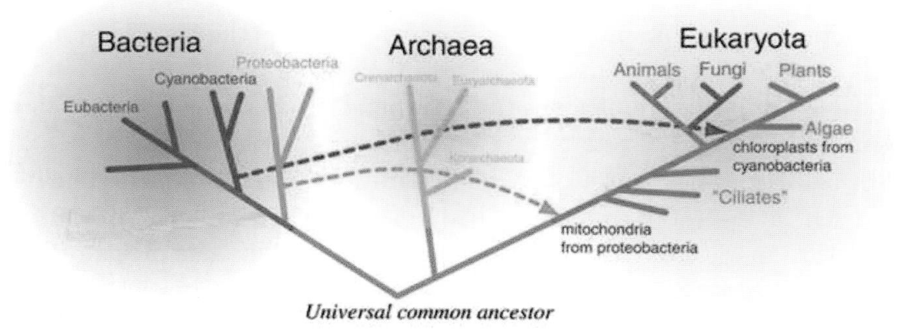

Fig. 9.3 The tree of life. All forms before the last common ancestor and most of those after, have become extinct. Permission: Alan Gishlich.

All known life on Earth, including one-celled organisms, require water in the liquid state, both within the organism itself, and in its environment. Oxygen, carbon dioxide, and nitrogen also are necessary. Through their leaves, plants absorb carbon dioxide from the atmosphere, using the carbon and exhaling the oxygen — a necessity of animal life. Plant and animal life thus have a symbiotic relationship, the one using the gas that the other expels. A small part of these gases was contained in the dust out of which the planet was formed. But the largest part is believed to have arrived during a heavy bombardment by icy planetesimals that swept through the inner solar system because of gravitational scattering by the giant planets, especially Jupiter. This occurred early in Earth's history, during the first 700 million years.

The great similarity of all life on Earth at the biological level is taken as evidence that all life *now* existing has a *last common ancestor* (Figure 9.3). From that ancestor, life has branched into three *domains*, Bacteria, Archaea and Eukarya. The Archaea are one-celled organisms with some genes like Bacteria and others like Eukarya (which includes animals). Many archaeans thrive in conditions that would kill other creatures: boiling water, super-salty pools, sulfur-spewing volcanic vents, acidic water and deep in Antarctic ice.

All life forms preceding the last common ancestor — and most forms afterward — suffered extinction. Only about one percent of life forms that appeared on Earth have survived to the present day.

The main chemicals of life are carbon, hydrogen, oxygen and nitrogen with small amounts of calcium, phosphorous and iron. There are three processes that distinguish life from the inanimate: (1) *biosynthesis* (the construction of organic molecules), (2) *reproduction* (the processes involved as cells make copies of themselves), and (3) *catabolism* (the process in which cells are broken into smaller ones). A transfer of energy is involved in each process.

The cell is the basic unit of all life from the multitude of one-celled creatures such as bacteria to multi-celled forms such as plants and animals. Humans are made from 10^{13} (10,000 billion) cells. The chemicals of life are contained within cells. Complex organic molecules of nucleic acids called RNA and DNA are central to protein synthesis and to the reproduction of cells.

9.3. Evolution of Life on Earth

> *Whilst this planet has been cycling on according to the fixed law of gravity, from so simple a beginning, forms most wonderful ... have been and are being evolved.*
>
> Charles Darwin (1809–1882)

Charles Darwin (Figure 9.4) sent his monumental work on evolution for publication in 1859 — *The Origin of Species by Means of Natural Selection.* It was the product of a large part of his lifetime of study following his youthful 5-year trip around the world in the sailing ship, the Beagle.

Darwin had taken a degree in theology from Cambridge University but he was much more interested in biology. The Reverend Professor John Stevens Henslow, a good-natured academic and clergyman taught Darwin much of his scientific technique, as well as arranging an unpaid place for Darwin — at the age of 22 — as a naturalist aboard the 90-foot sailing ship, H.M.S. Beagle. The ship was bound for a 5-year voyage around the world and his stop at the Galápagos Islands awakened him to possible causes for the variation in finches that he observed on the different islands of that group. He gathered specimens of animal and plant life throughout the voyage, sending preserved specimens back to England from those rare ports of call where a ship bound for the homeland could be found. In 1837, one year after he returned from the voyage on the Beagle, he made detailed notes on the idea of evolution by means of natural selection. At that time Alfred Wallace, who was to arrive years later at a similar theory, was only 14.

There are several factors that lead to the gradual evolution of a species. Random variations occur between parents and their offspring. Many are damaging but some give an advantage. Those with the advantage may pass

Fig. 9.4 Charles Darwin (1809–1882) at the age of 7 years and as an old man.

it on to their offspring. Mutations caused by the habitat such as cosmic rays, changes in climate, and competition from other species are also factors that play a role in production and selection of favorable mutations. In this way members of the species with the favorable variation may supplant others of the same species, and *that* species may supplant other species altogether that are competing for food and reproduction in the same environmental niche. It is thought that over the Earth's history, about 99% of species have become extinct. Natural selection of favored mutations can be observed in laboratory experiments using organisms with a short reproduction rate such as bacteria and fruit flies. On longer timescales, it is the fossil record that provides evidence of evolution (Figure 9.5).

9.4. Earliest Life on Earth

The Sun and solar system formed about 4.5 billion years ago and Earth was subjected to a heavy bombardment from space by rock and ice planetesimals for the first 700 million years. The first life arose before this according to the

Fig. 9.5 A trilobite fossil (ptychopqriida, redlichida) from the Cambrian period about 500 million years ago. Credit: Framlingham Sir Robert Hitcham's CEVAP School; website address (www.hitchams.suffolk.sch.uk).

fossil record of bacteria in Australia; the early maelstrom from space utterly destroyed it. New life arose soon after the bombardment: the evidence is found in rocks from Greenland that contain the isotope of carbon, ^{12}C, which is produced by plant life (photosynthesis) in higher than normal abundance in the environment. There is also less controversial fossil evidence of early life from 3.4 billion years ago in South Africa. This much is known.

One may speculate on where life on the planet first arose. The oceans, near the surface where sunlight could penetrate, or in shallow tidal pools seem plausible locales. Other molecular biologists speculate differently. The tree of life, Figure 9.3, shows that, of Archaea, a large group are Thermophiles that flourish near hydrothermal vents, suggesting that life originated deep in the Earth. As soon as self-replicating carbon-based bimolecular systems emerged, they rapidly began to claim the environmental resources and there would have been negligible chance for any other type of system to survive.

Fig. 9.6 An impact crater in the Arizona desert made by a meteor is a mile wide, 570 feet deep with a rim 150 feet high. It landed an estimated 50,000 years ago and had an explosive force estimated to be 2.5 million tons of TNT. It was made of nickel-iron, weighed 300,000 tons and landed with a speed of 12 km/sec. This crater is very small compared to the one buried in the jungles of the Yucatan.

9.5. Mass Extinctions

At certain times over the 4.5 billion year life of the Earth, slow changes in Earth's climate or cataclysmic events have occurred which caused the extinction of some species and favored the ascendance of others. The impact of the meteor — some 7–10 miles across — that struck the Yucatan peninsula of Mexico 65 million years ago is believed to have extinguished the dinosaurs and 70% of species living then, and perhaps paved the way for the ascendance of mammals. The crater, known by the Indian name — Chicxulub — had been known of only since the 1950s, when it was discovered during oil drilling. It was buried under more than a half a mile of limestone deposits. The impact produced fires, acid rain and tsunami-like destructive waves. The collision gouged a crater 125 miles in diameter, nearly eight miles deep, and sent an enormous quantity of rock, dirt, and debris spinning into the Earth's atmosphere. The material blocked sunlight, causing extreme changes in the Earth's climate worldwide for a period of several years. Deposits that settled from the atmosphere as far away as Colorado are seen in Figure 9.7.

Fig. 9.7 In the Raton Basin in Colorado, the boundary between the Cretaceous-Tertiary periods of 65 million years ago correspond to the huge meteor impact in the Yucatan. The light gray in the middle of the photograph (marked with the red knife) is composed of debris from that impact. Coal deposited in the Tertiary Period, after dinosaurs disappeared, is deposited on top of the C-T boundary layers. Photo credit: David A. Kring.

9.6. Northern Lights (Aurora Borealis)

Northern lights originate from large explosions and flares on the Sun's surface. During these eruptions, enormous numbers of electrically charged solar particles (electrons and protons) are cast out of the Sun and into space. These particles travel with speeds up to a million km per hour. But even with such speeds it takes them two to three days to reach Earth. Those that approach Earth are guided toward its two poles by the Earth's magnetic field and create aurora over both poles.

On their way down towards the Earth's geomagnetic poles, the solar particles are stopped by the atmosphere, which acts as an effective shield against these particles; *they would otherwise destroy life on Earth.* When the solar particles collide with the atmospheric gases, the collision energy is transformed to photons of light. A large number of such collisions create

Fig. 9.8 Aurora Borealis produced by electrons and protons streaming from the Sun producing billions of photons of light when they strike Earth's atmosphere. Were the charged particles not stopped by the Earth's sufficiently thick atmosphere, they would destroy life. Credit: Northern Lights Photo of the Year 2001 by Phil Hoffman, Canada.

an aurora — a light show as in Figure 9.8 that may move across the sky or disappear and reappear at another place in the blink of an eye. For an observer to actually see the aurora with the naked eye, about 100 million photons are required. The aurora does not actually reach Earth's surface; it extends from about 40 to 600 miles above the surface.

The Earth's magnetic field is distorted by particles streaming from the Sun's hot surface. It is pushed toward the Earth on the side facing the Sun, and is formed into a long plume pointing away from Earth on the other side (Figure 9.9). The magnetosphere extends into the vacuum of space from approximately 50 to 37,000 miles on the side toward the Sun, and trails out to a distance of more than 50 Earth radii from the Sun (Figure 9.9).

9.7. Habitable Zone around Stars

The *habitable zone* around a star refers to the range of distances from the star in which liquid water would exist on planets. On Earth, life exists at

Fig. 9.9 The magnetic field of the Earth is flattened on the illuminated side of the Earth by the stream of charged particles radiating from the Sun, and pushed into a long plume on the other side. Credit: NASA.

temperatures between −18 and 123°C. As well as water, a habitable planet must be massive enough that gravity retains an atmosphere containing the gases necessary for life such as oxygen and carbon dioxide; otherwise these gases would escape into space. The atmosphere must be sufficiently thick that its pressure on the surface of the planet maintains water in its *liquid state*, as is necessary for life on Earth.

Additionally, for a planet to be habitable, it must not be too close to the Sun. Otherwise the cloud cover would be too thick and the planet's temperature will rise because of trapped infrared radiation. Nor should its orbit be too elliptical; otherwise temperature variations over a year would be excessive. These are necessary conditions for a planet to maintain carbon-oxygen liquid-water based life as on Earth. (One cannot rule out the possibility of life based on an entirely different chemistry. But such is unknown.) The interior of the Earth to a depth of 5 km is not so hot as to be devoid of carbon-oxygen-liquid water based life. However, only unicellular life exists to such depths.

Mercury is closer to the Sun than Earth. As it rotates, its temperature varies between 400°C during the day and −180°C at night. However, because temperatures are higher toward the core, primitive forms could exist to some depth.

Venus' surface temperature is above 450°C; this is sustained by an enormously thick cloud cover that creates a greenhouse effect such that its temperature is almost constant.

Mars has abundant water that may exist in liquid form under its surface. However, it is too hot during the day (160°C) and too cold at night (−180°C) to have surface life. But because temperatures are higher toward the core, primitive forms could exist to some depth.

Chapter 10

Other Planets and Their Moons

The contemplation of celestial things will make a man both speak and think more sublimely and magnificently when he descends to human affairs.

Marcus Tullius Cicero, 106–43 B.C.

Fig. 10.1 The four *terrestrial* planets of the solar system shown here — Mercury, Venus, Earth and Mars — are closer to the Sun than the *gaseous* planets, Jupiter, Saturn, Uranus, Neptune and the icy rocky planet, Pluto. Credit: NASA.

10.1. Jupiter

Zeus (Jupiter) was the king of the Greek gods, ruler of Olympus and son of Cronus (Saturn). Jupiter has four large moons, called the Galilean moons (but named by Galileo the Medicean moons after his patrons) and about 63 others (as of February 2004); most likely there are smaller ones not yet seen. The discovery of minor moons continues to this day. With its many moons and several rings, Jupiter is a mini-solar system. It is the fourth brightest

Fig. 10.2 Jupiter, as photographed by the unmanned spacecraft Voyager 1 (see page 202 for details of these space flights) with its largest Galilean moon, Ganymede, visible near the bottom. This moon is more massive than the planet Mercury. Jupiter appears striped. These stripes are dark belts and light zones created by strong east-west winds in Jupiter's upper atmosphere. Within these belts and zones are storm systems that have raged for years. A prominent feature of Jupiter is the Great Red Spot, first observed by Galileo four centuries ago, which is a storm larger than Earth. Credit: NASA Voyager 1, JPL.

object in the sky (after the Sun, the Moon and Venus) and it is clearly visible in the night sky. Jupiter takes 12 years to orbit the Sun but its day is only about 10 hours. Its diameter is 11 times greater than Earth's.

If Jupiter (Figure 10.2) had been between about 100 times more massive, the great weight of its exterior, pressing on the inner region, would have ignited thermonuclear reactions as in our Sun, converting gravitational energy to heat and pressure: it would have shone — it would have been a star!

Jupiter has two-thirds of the combined mass of all the other planets in the solar system. Its rocky core alone is 10–15 times the mass of the Earth, and altogether it is 320 times more massive than Earth. Otherwise, Jupiter is a gas planet, made 75% of hydrogen and 25% of helium as measured in mass, with traces of methane, water and ammonia. It has a rock core. The hydrogen and helium are primordial elements, made in the first few minutes in the life of the universe and they are in the same primordial proportions.[a] The others, which by comparison are called trace elements because they amount to so little, were made much later after the first stars had formed around 200,000 million years after the birth of the universe. Saturn has a similar composition to Jupiter.

Above the rock core of Jupiter lies the main bulk of the planet in the form of liquid metallic hydrogen. This exotic form of the most common of elements is possible only at pressures exceeding 600 million pounds per square inch, as in the interior of Jupiter and Saturn. Liquid metallic hydrogen consists of ionized protons and electrons (like the interior of the Sun but at a far lower temperature). It is an electrical conductor and the source of Jupiter's magnetic field. This layer probably also contains some helium and traces of various "ices."

Jupiter's rotation is effected by its moons in a manner similar to the way Earth has effected the rotation of our Moon. The process responsible for the gradually slowing of Jupiter's rotation is due to the tidal drag produced by its Galilean satellites (see page 144). Also, the same tidal forces are changing the orbits of Jupiter's moons, slowly forcing them farther away. At the same time, the constantly changing shape produced by the drag, liquefies the interior of the inner moons into molten lava which gives rise to colorful plumes of gaseous material on the moon, Io, as in Figures 10.3, 10.4 and 10.5.

Jupiter radiates more energy into space than it receives from the Sun. This heat is probably produced from the slow gravitational compression[b] of the planet so that its core may be as hot as 20,000°C. This great interior heat liquefies the interior and generates convective currents deep within the planet.

[a]Although helium is also made in stars by burning hydrogen, the helium is further burned to form heavier elements, and so does not disturb the primordial abundances.
[b]Conversion of gravitational energy into heat, which is a measure of the kinetic energy of motion of the atoms and molecules.

Fig. 10.3 Two sulfurous eruptions are visible on Jupiter's volcanic moon, Io, in this color composite. Io is the most volcanically active body in the solar system. On the left, over Io's limb, a new bluish plume rises about 86 miles above the surface of a volcanic caldera known as Pillan Patera. In the middle of the image, near the night/day shadow line, the ring shaped Prometheus plume is seen rising 45 miles above Io, while casting a shadow to the right of the volcanic vent. Named for the Greek god who gave mortals fire, the Prometheus plume is visible in every image ever made of the region dating back to the Voyager flybys of 1979 — suggesting that this plume has been continuously active for at least 18 years. Image recorded on June 28, 1997 at a distance of 372,000 miles. Credit: JPL, NASA.

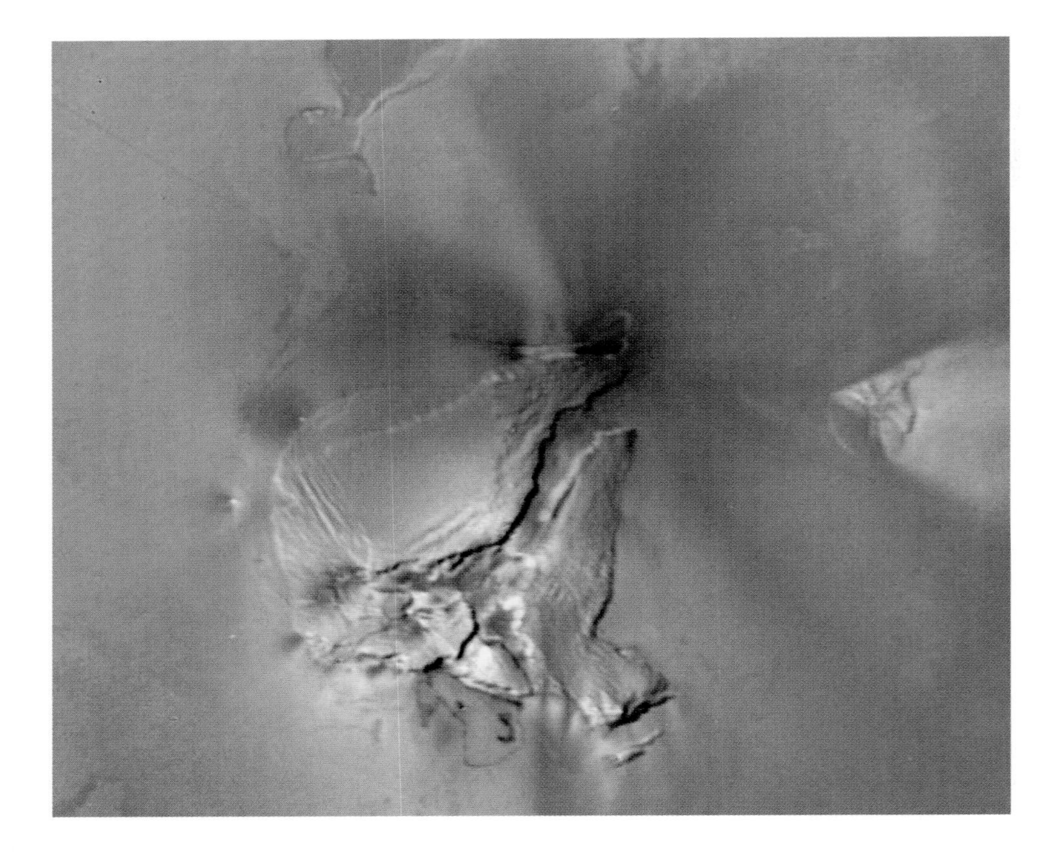

Fig. 10.4 Vent source of the Pele plume on Io (one of Jupiter's moons), observed by Voyager in March 1979. The plume extends 500 km (312.5 miles) in all directions. The eruption source is the small dark spot just above the center of the frame. Credit: NASA.

As Jupiter rotates, a giant magnetic field is generated in its electrically conducting liquid interior in the same way the Earth's field is generated (see page 143). Within Jupiter's magnetosphere — the area in which magnetic field lines encircle the planet from pole to pole — there are so many charged particles as to make the inner portions of Jupiter's magnetosphere the most deadly radiation environment of any of the planets, both for humans and for electronic equipment (important for space exploration).

The tail of Jupiter's magnetic field stretches behind the planet as the solar wind rushes past; it extends billion of miles behind Jupiter — as far as Saturn's orbit. Jupiter's rings and moons are embedded in an intense radiation belt of electrons and ions trapped in this magnetic field.

It is interesting to speculate on the reason that the gaseous outer part

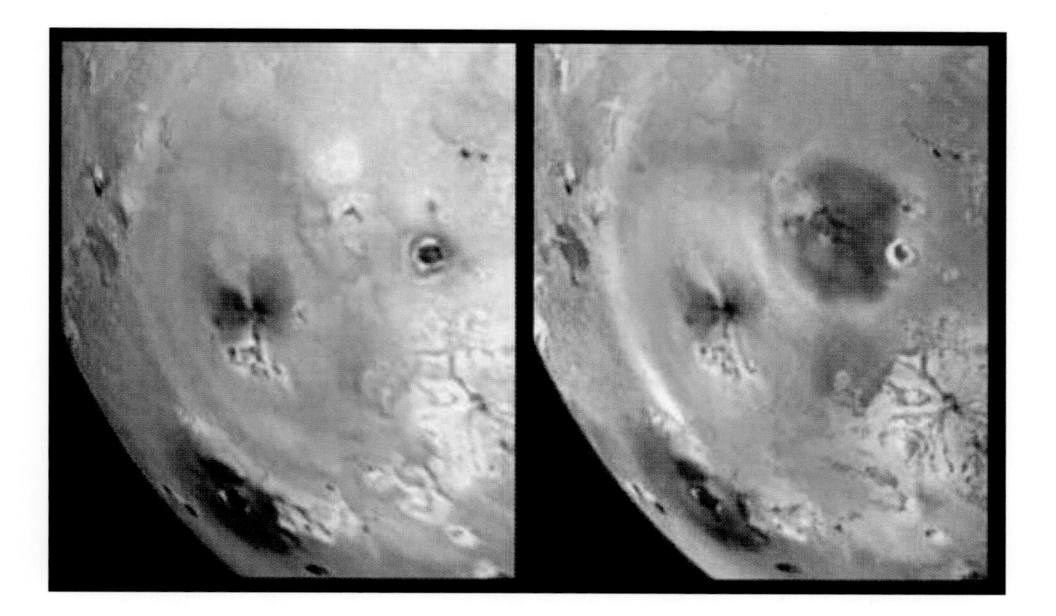

Fig. 10.5 The three red patches on the surface of Jupiter's moon, Io, are volcanic plumes. The annular red plume deposit of sulpher dioxide frost is evident in these paired images of the Pele hotspot. The image on the left was taken by Galileo Orbiter on April 4, 1999 and that on the right on September 19, 1999. In less than six months an eruption occurred on Pillan Patera to create the dark region visible to the right of Pele in the September image. The dark spot is probably composed of silicate fall-back from an eruption. Credit: NASA.

of Jupiter contains hydrogen and helium in the same proportion with which they appeared in the first few minutes of the universe. Perhaps it is because the gravity of the very massive rocky core, some 10–15 times Earth's mass, attracted the interstellar gas which would have contained these elements in the primordial abundance.

10.2. Moons of Jupiter

Jupiter has many moons, most of them small. Because of their size, 11 of them were only recently discovered at the University of Hawaii's 88-inch telescope on Mauna Kea. The newly discovered moons have diameters from 3 to 8 km, and most of them are in very large orbits, some strongly inclined from the orbital plane that Jupiter makes with the Sun. Two of them actually circle the big planet in the opposite direction along with 5 other previously known satellites that are in retrograde orbit.

Galileo (1564–1642) discovered the four largest moons of Jupiter — Io,

Europa, Ganymede, and Callisto, in 1610, which he named the Medician moons after his patrons, though they are now referred to as the Galilean moons. His discovery followed very shortly after the invention of the telescope by a Dutch spectacle maker, who used the low-power lenses of his trade, and which Galileo greatly improved. With his discovery of Jupiter's moons and the phases of Venus, Galileo became convinced that the Earth was not the center of the world (see page 154).

The larger moons of Jupiter are spheres like our moon, made so by gravity, but compared to Jupiter, they are very small. Ganymede is the largest moon in the solar system, almost as large as Mercury. The spacecraft Galileo, during its first flyby, discovered that Ganymede has its own magnetic field embedded inside Jupiter's huge one. This probably means that the interior is molten iron which generates a field in a similar fashion to the Earth's as a result of the motion of conducting material in the interior.

Io is the innermost of Jupiter's four large moons and the third largest of the four discovered by Galileo. Io has a diameter very close to that of our moon. It is intensely volcanic. Because of its non-circular orbit, the gravitational pull of Jupiter constantly changes Io's shape causing its interior to be tidally heated (page 144). Consequently, molten lava spews to its surface at frequent intervals; Io is dotted by hundreds of volcanic centers, about 70 of which are active. It is the only body outside the Earth to exhibit active volcanism on a massive scale.

Craters created by meteors and other small bodies scar the surface of our Moon. Such impact craters are absent from Jupiter's moon Io because of the vigorous volcanic activity which can be seen in Figure 10.3 taken in 1979. Older structures would have been covered by lava, sulfur, and pyroclastic materials, which have formed the dimpled and variegated crust that it exhibits today. Io's very frequent volcanic eruptions can be gauged in Figure 10.5 where in a six-month interval, new eruptions can be detected.

On Earth, the source of heat that produces volcanic activity comes from energy released by the decay of radioactive materials within the interior, as well as from heat left over from Earth's formation at the time it was heated by the impacts of planetesimals. (See Section 7.7.) But Io, one of Jupiter's moons, is too small to have left over impact-induced heat, and radioactive decay could not generate the tremendous energy required to power all of the volcanic activity that exists on that moon. The answer is tidal heating.

We have described the phenomenon of tidal heating of our Moon as it orbits Earth (page 144). So too the four largest moons of Jupiter — Io, Europa, Ganymede, and Callisto — are tidally heated as they orbit the big

Fig. 10.6 Mars, the red planet as seen from Earth. The image was taken during the middle of summer in the Southern Hemisphere. During this season the Sun shines continuously on the southern polar ice cap, causing the cap to shrink in size. The orange streaks are indications of dust activity over the polar cap. The cap is made of carbon dioxide ice and water ice, but only carbon dioxide ice is seen in this image. Winter envelops the planet in the north. Credit: NASA, J. Bell (Cornell U.) and M. Wolff (SSI).

planet. Those moons range in size from Earth's Moon to as large as the planet Mercury. Because of tidal heating, water exists in ice form on the latter three but may be liquid in their warmer interiors.

Io is the most tidally heated of the four Galilean moons and it is also heated by energy released from the radioactive decay of long-lived isotopes in its interior. Io's surface has no landforms resembling impact craters; this is because volcanic activity is so great (Figure 10.5) that Io is resurfaced with new lava deposits from its interior more rapidly than the impact of comets and asteroids from outer space can create large craters.

10.3. Mars

Romulus and Remus were the twin sons of Mars. As babies, they were cast into the Tiber River. The brothers were miraculously rescued by a she-wolf who reared them with her cubs underneath a fig tree. After a few years they were found by a shepherd who took the brothers to his home. When they reached maturity they built a settlement on the Palatine Hill. During a quarrel, Romulus slew Remus and became the sole ruler of the new Rome. To enlarge his empire, he allowed exiles, refugees, and runaway slaves to populate the area and to solve the shortage of women he stole the Sabine women whom he invited to a festival. Upon his death he was taken to the heavens by his father Mars.

Homer, *ancient folk tale*

Although Mars is a small planet — its radius is just a little over half of Earth's — its scenery is on a scale that dwarfs Mount Everest and the Grand Canyon. The Grand Canyon is a mere 280 miles long, and up to 18 miles wide, whereas running roughly along the equator of Mars, there is a split in the crust that is 2,400 miles long (about a fifth of the distance around the whole of Mars), and up to 360 miles wide and 4 miles deep. The Hellas Basin in the southern hemisphere, which is a crater created by the impact of a meteor or asteroid, is 1500 miles in diameter and 6 miles deep.

Mars has the highest volcano in the solar system, Olympus Mons (Figure 10.7) which stands 17 miles above the surrounding plain: Mount Everest is only one-third as high. Mars also shows evidence for the most catastrophic floods. These floods have carved large channels that drain into the northern plains, suggesting that an ancient ocean covered most of the northern hemisphere. Valley networks that crisscross the southern highlands suggest that rivers once flowed on Mars. However, water disappeared about 3.8 billion years ago when the solar system was quite young (present age, 4.5 billion years). Much of its atmosphere and all of any surface water has long since vanished. Today, atmospheric pressure at ground level is only about one-hundredth that on Earth.

10.4. Moons of Mars

Earth's Moon is an exception among the moons in the solar system; it is the largest compared to its planet and, like many other massive moons, it is spherical, made so by gravity. While Earth's Moon was formed at the

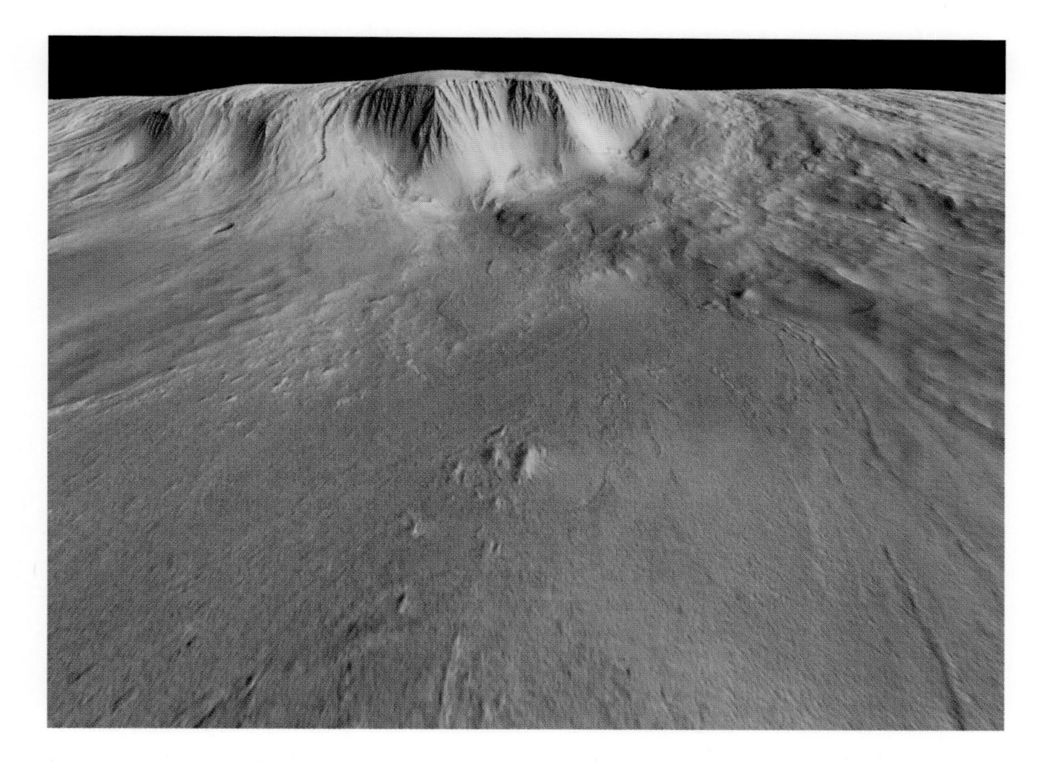

Fig. 10.7 The ledge of the giant Olympus Mons on Mars, shows only a part of the escarpment which is 4 miles high. The peak of the volcano, lying far in the background, is more than 12 miles above that. The diameter of the volcano is 374 miles. Credit: European Space Agency (ESA).

same time as the Earth and other planets, the two moons of Mars were likely captured asteroids, attracted by Mars' gravity. They would be much older than the solar system. These moons are small and irregular in shape. Phobos is a mere 17 by 12 miles and Deimos is still smaller, 10 by 8 miles. Mars itself, has been known since antiquity, but its moons were discovered as recently as 1877.

Spacecraft sent to Mars have photographed its moons, one of which is shown in Figure 10.8. Phobos has craters of all sizes on it; it also has long grooves radiating from its largest crater. Deimos has a few small craters. A fine dust covers its surface, but it also has boulders on it — some as large as a house. The marred surfaces and irregular shapes of Mars' small moons are likely consequences of impacts in space when the moons were still asteroids. Such impacts have probably melted the moons at various times through their long history. Unlike the planet itself, Mars' moons have not been known since antiquity. Rather, they were discovered in 1877.

Fig. 10.8 The larger of Mars' two tiny moons, Phobos measures only 17 miles along its greater length, and was not known until relatively recently. The pits and scars on Phobos are partially filled in, probably from a time when it roamed the solar system before its capture by Mars. At that time it would have been heated and melted by impacts with other interstellar debris. Photo credit: NASA.

10.5. The Ringed Planet, Saturn

> *Sani was identified with the planet Saturn. Sani's parents were the Sun god Surya and Chhaya whose name means shade. Surya's wife could not tolerate the intense heat produced by her husband the Sun god. To protect herself, she sent a shade to her husband in the form of a mistress whose name was Chhaya. Sani was known as the evil-eyed one because his glance was extremely powerful and could burn anything instantly.*
>
> Hindu myth

Saturn is 10 times further from the Sun than Earth and is the sixth planet in the solar system. It is the second largest planet; its equatorial diameter is almost 10 times larger than Earth and its mass is 100 times greater. Much of what is known of Saturn was learned through the Voyager explorations in 1980–81. It is flattened at the poles and stretched at the equator because of

Fig. 10.9 Saturn's rings are a great mystery — how they were formed, of what, and the forces that drive them are unknown. Maybe Saturn's rings are fragments of a moon that ventured too close to the planet early in its formation and was broken up by the differential force of Saturn's powerful gravitational filed. Two of Saturn's moons, Rhea and Dione are visible as tiny dots at the bottom of the figure. The unmanned spacecraft, Voyager, made its first visit to Saturn in 1980. Credit: JPL, NASA.

Fig. 10.10 This is a color enhanced image of Saturn's ring system. What are they made of? The bluer area appears to be rich in water ice. The redder areas appear to be richer in some sort of dirt. This and other images show that the inner rings have more dirt than the outer rings. The thin red band in the otherwise blue A ring is the Encke Gap. The exact composition of dirt remains unknown. Credit: NASA and JPL.

its very fast rotation. Its day is a mere 10.5 hours but it takes 30 years to orbit the Sun. Saturn is less dense than water. Its atmosphere is composed mostly of hydrogen with small amounts of helium and methane.

A wind blows at high speeds on Saturn. Near the equator, it reaches velocities of 1,760 km an hour (1,100 mph). The wind blows mostly in an easterly direction. The strongest winds are found near the equator and the velocity falls off uniformly at higher latitudes. At latitudes greater than 35 degrees, winds alternate in direction as the latitude increases.

The rings of Saturn, Figure 10.9, make it one of the most beautiful objects in the solar system. The rings were first dimly observed by Galileo with his self-made telescope soon after its invention in Holland. "I have observed the highest planet [Saturn] to be tripled-bodied. This is to say that to my very great amazement Saturn was seen to me to be not a single star, but three together, which almost touch each other." He had seen two lobes, one on each side of Saturn, but his telescope did not have the power to reveal them as the extremes of the rings around Saturn.

The rings are split into a number of different parts that are divided by gaps (Figure 10.10). An Italian-born French astronomer, Giovanni Cassini discovered this division in 1675. Cassini was the first director of the Royal Observatory in Paris. He discovered four of Saturn's moons and the major gap in its ring. Besides its rings, Saturn has at least 33 moons.

10.6. Venus

The ancient Mayans used the doorways and windows of their buildings as astronomical sightings, especially for the planet Venus. At Uxmal (Mexico), all buildings are aligned in the same direction.

The planet Venus, sometimes referred to as the evening and the morning star by the ancients, was thought by Greeks and Romans alike to be the goddess of love and beauty; she was known to the Greeks as Aphrodite. Julius Caesar and the emperor Augustus claimed her as the ancestor of their Julian family.

Galileo was not the first to see Venus, but he was the first to understand that it was a planet. In his time it was Church doctrine — which accepted Aristotle as its authority — that the nearby heavenly bodies circled Earth (see Section 8.5). It was Aristarchus of Samos (310–230 B.C.), a mathematician and astronomer, who correctly maintained that the Earth and other planets orbited the Sun. But his views seemed to contradict the daily

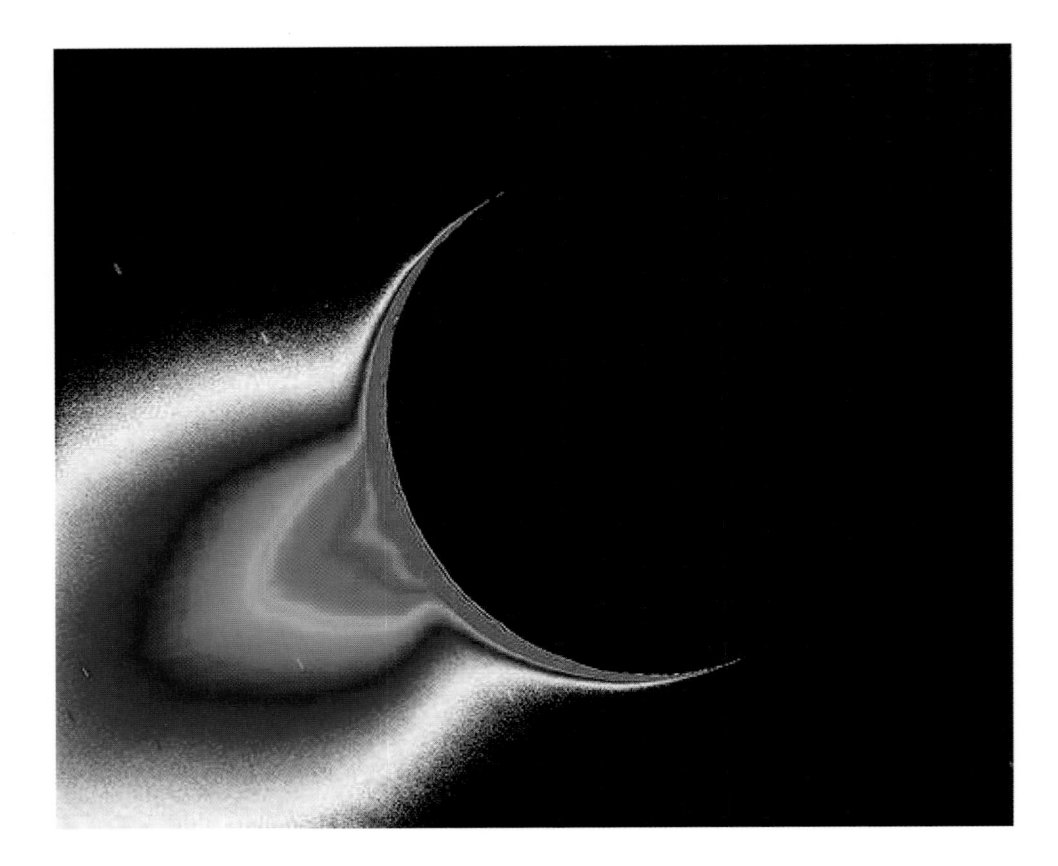

Fig. 10.11 Space mission Cassini obtained the images of Saturn's moon Enceladus backlit by the Sun and showing a fountain-like sources of the fine spray of material that towers over the south polar region. The greatly enhanced and colorized image shows the enormous extent of the fainter, larger-scale component of the plume. Credit: NASA, the European Space Agency, and the Italian Space Agency.

observation that the Sun rose in the east and set in the west: To the common man, this meant that the Sun revolved about Earth and that Earth was at the center of the universe.

However, some 400 hundred years later, Aristotle had some influence on Ptolemy (100–178 A.D.), who invented an elaborate system of movements of the planets to account for their motions, which appeared very complicated in this view. Thus we have Ptolemy's system of cycles and epicycles.

Ptolemy, a Greek astronomer living in Alexandria, believed that a sphere enclosed the Universe with the Earth at the center. The stars were fixed on a great enclosing membrane. According to his scheme the planets described circular orbits whose centers rotated in another circle, around a point that deviates from the center of the Earth. By varying the speed of the planets and diameter of the various circles this gave an acceptable description of the planetary movements.

In fact Ptolemy's artificial system was able to account for the planetary motions over long periods of time, and stood as an obstacle for further progress in astronomy for centuries. Indeed, until Copernicus in the 16th century recorded his detailed observations which so much influenced Galileo and the course of his life, Ptolemy's artificial system was the accepted account of planetary motion.

Galileo used his telescope to observe that Venus went through phases of illumination by the Sun, being illuminated on one side and later on the other. He concluded, therefore, that Venus circled the Sun, not the Earth. This contradicted the Church belief that Earth was the center of the world. It was among the most important observations in human history, for it provided the first conclusive proof that was consistent with the Copernican system and contradicted the Ptolemaic system. Galileo's observation was a first step in the long journey to our present understanding of the universe.

Venus, unlike any other planet in the solar system, rotates about its axis backwards as compared to the others. It is the second planet from the Sun, and is Earth's closest neighbor in the solar system. Venus is the brightest object in the sky after the Sun and the Moon, and sometimes looks like a bright star in the morning or evening sky. The planet is slightly smaller than Earth, and its interior is similar to Earth.

Venus is shrouded in thick clouds that strongly reflect sunlight. The atmosphere is composed of carbon dioxide, and clouds of sulfuric acid. Its surface is scorched with temperatures of more than 450°C — high enough to melt lead. But the heat that is radiated outward is trapped by the dense atmosphere and cannot escape into space. This effect is caused by the heavy atmosphere — 90 times heavier than our own — about the pressure at 3000 feet below sea level.

Measurements made by probes that traveled through the atmosphere, have shown that the temperature remains nearly constant through the long dark night. Thus there are neither significant seasons nor daily temperature changes in the atmosphere.

Space missions to Venus have revealed over 1600 major volcanoes, mountains, large highland terrains, and vast lava plains. Flows from volcanoes have produced long sinuous channels extending for miles, with one extending nearly 4,300 miles.

The surface of Venus is peppered with large impact craters created by

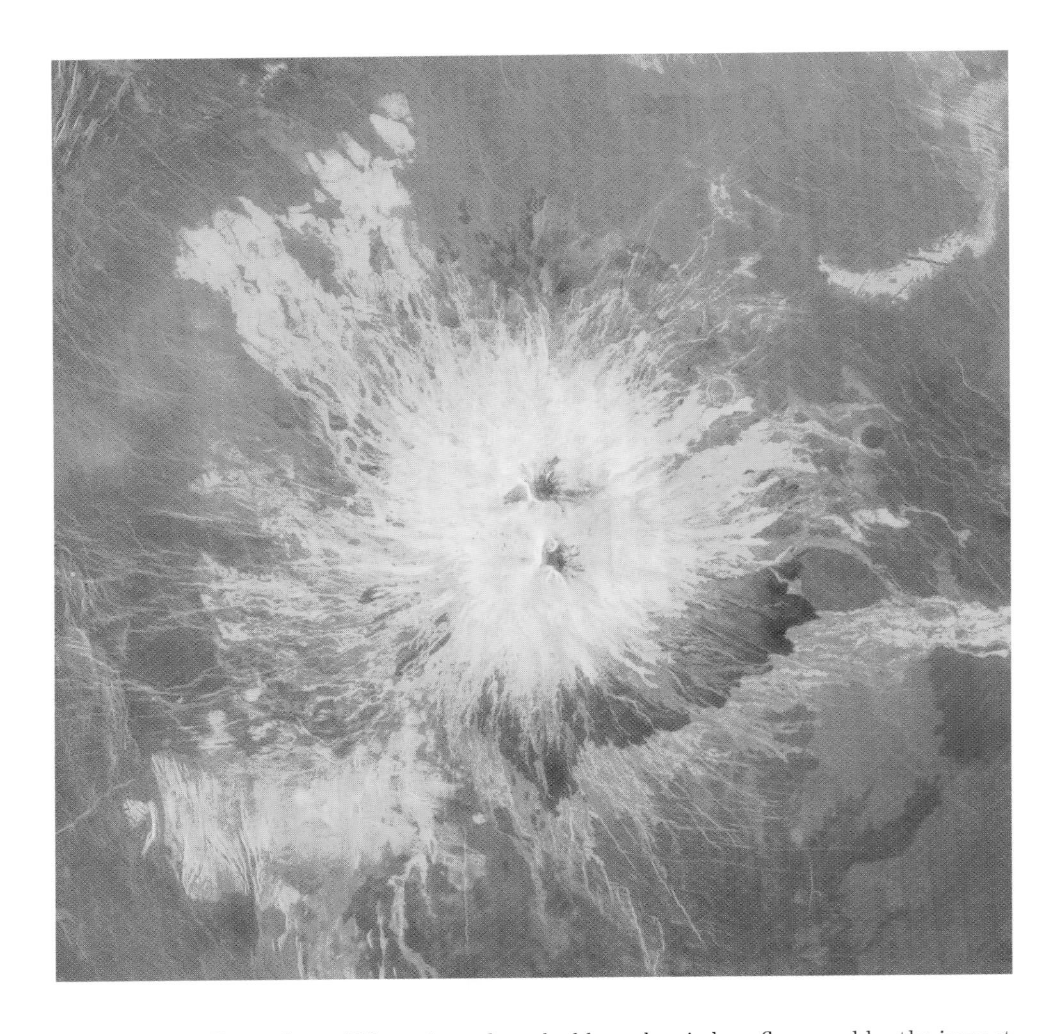

Fig. 10.12 The surface of Venus is pockmarked by volcanic lava flows and by the impact of meteors and asteroids. Shown here is a false-color image of the volcano Sapas Mons. The area shown is approximately 650 km (404 miles) on one side. Sapas Mons measures about 400 km (248 miles) across. Courtesy: NASA.

large meteors or asteroids and by volcano lava flows such as those seen in Figure 10.12. The impact craters are large; smaller objects would have burned up in the thick atmosphere of Venus before reaching the surface. It may seem strange that the impact craters all appear to have been made by vertically incoming meteors with no glancing blows. However, this conclusion would be wrong; it has been found by observations and *experiments* on Earth that even quite oblique impacts produce round craters (see Figure 9.6).

Fig. 10.13 Mercury, closest planet to the Sun. Colors are enhanced in this Mariner Orbiter photo. Its surface is pitted by craters. Credit: NASA.

10.7. Mercury

The Romans named Mercury after the fleet-footed messenger of the gods, because it seemed to move in the sky more quickly than any other planet. It is the closest planet to the Sun, and smallest planet in the solar system. It has a mass of only about 1/20 of Earth's.

In the 1880s Giovanni Schiaparelli believed that Mercury was tidally locked to the Sun in the same way our Moon always faces the same way to Earth (see page 144). However, radio emissions from the planet detected by astronomers in the 1960s found that the dark side was too warm to be forever facing away from the Sun. But Mercury does rotate very slowly, only once in every 59 (Earth) days. So it is not yet tidally locked. Because of its slow rotation, it becomes extremely hot (400°C) during its long day and cold (−150°C) during its long night. In the distant past it must have rotated much faster — about 8 hours from sunrise to sunset. But because of its tidal coupling to the Sun its days have stretched enormously to 176 Earth days. Its year is only 88 Earth days. Mercury rotates 1.5 times during each orbit of the Sun.

Mercury has a very weak magnetic field; only about one percent of Earth's as found by NASA's orbiter Mariner in the 1970s. Its entire surface is therefore unprotected against bombardment by high-energy electrons and protons from the solar surface, whereas the Earth's field deflects most of these particles to the poles.

10.8. Uranus

Uranus was born from Gaea, the Earth, and he became the god of the sky. His rains fertilized the Earth and gave birth to many kinds of human-like creatures, one of which were the Cyclopes and another were the Titans. Uranus tried to confine the Cyclopes to the depths of the Earth, a place known as Tartarus.

Greek myth

Uranus is a gas giant, the seventh planet from the Sun and the third largest by diameter and fourth largest by mass. It is named after Uranus, the Greek god of the sky, and progenitor of the other gods. Uranus was not known as a planet in ancient times, although it had been observed but was always mistakenly identified as a star. In the Chinese, Japanese, Korean, and Vietnamese languages, the planet's name is literally translated as the sky king star. The earliest recorded sighting in the west was in 1690 when John Flamsteed catalogued it as 34 Tauri. Uranus has 27 known moons. It has a faint planetary ring system that was discovered as recently as 1977. As of 2005, 13 rings had been identified.

The planet is composed primarily of rocks and various ices, with only about 15 percent hydrogen and a little helium. Uranus (like Neptune) is in many ways similar to the cores of Jupiter and Saturn without the massive liquid metallic hydrogen envelope. It appears that Uranus does not have a rocky core like Jupiter and Saturn but rather that its material is more or less uniformly distributed. The surface temperature on Uranus's cloud cover is approximately −218°C.

Unlike any of the other planets, the rotation axis of Uranus is tilted at 98 degrees to its orbit. Consequently, for part of its orbit, one pole faces the Sun continually while the other pole faces away. Between these extremes the Sun rises and sets around the equator normally. Uranus' extreme axial tilt probably originated during the formation of the planet when it may have collided with another protoplanet.

10.9. Neptune

Neptune is the outermost of the gas giants. Its mass is about 17 times greater than Earth. Neptune has eight moons, six of which were found by NASA's unmanned Voyager. A day on Neptune is more than 16 hours. The inner two-thirds of Neptune is composed of a mixture of molten rock, water, liquid ammonia and methane. The outer third is a mixture of heated gases composed of hydrogen, helium, water and methane. Methane gives Neptune its blue cloud color.

Neptune was discovered on September 23, 1846 by Johann Gottfried Galle, of the Berlin Observatory, and Louis d'Arrest, an astronomy student, through mathematical predictions made by Urbain Jean Joseph Le Verrier.

The discovery was not by accident. It had been proposed that a planet beyond Uranus could account for irregularities in that planet's orbit. Independently, two astronomers, John Couch Adams in England and Urbain Jean Joseph Le Verrier in France, calculated the position of this yet unknown planet by its effect on other planets. The British astronomer James Challis, using Adams' predictions, observed Neptune on the night of August 4, 1846, but failed to compare his observations with those of the previous night and did not recognize the planet. On September 23, 1846, the planet was finally found on the first try by the German astronomer Johann Galle using Le Verrier's predictions.

Neptune is a dynamic planet with several large, dark spots reminiscent of Jupiter's hurricane-like storms. The largest spot, known as the Great Dark Spot, is about the size of the Earth and is similar to the Great Red Spot on Jupiter. The strongest winds on any planet were measured on Neptune. Most of the winds there blow westward, opposite to the rotation of the planet. Near the Great Dark Spot, winds blow up to 2,000 km (1,200 miles) an hour.

Neptune has a set of four rings that are narrow and very faint. The rings are made up of dust particles thought to have been made by small meteorites smashing into Neptune's moons. The magnetic field of Neptune, like that of Uranus, is highly tilted at 47 degrees from the rotation axis and offset from the physical center by at least 0.55 radii (about 13,500 km or 8,500 miles).

10.10. Pluto (a Dwarf Planet)

Pluto is so small and far away that it was not discovered until 1930. It *was* nominally the last planet in the solar system, only about two-thirds the size of our Moon. It is often referred to as a minor planet; there are many other minor planets beyond Neptune, more than ten thousand. It has recently been demoted to a *dwarf* planet by a board of astronomers.

Several astronomers interpreted irregularities in the orbits of Uranus and Neptune as being caused by a more distant planet. Among these astronomers was the American Percival Lowell who is credited with the successful prediction of Pluto's orbit. Lowell also started the search for the planet, which was ultimately found in 1930 by his successor at the Lowell Observatory in Arizona, Clyde Tombaugh. He made the discovery by comparing a photographic plate taken on the night of January 23, 1930 with two others taken in the same month. Because Pluto is so small and so distant from the Sun, the image of the planet is extremely faint.

Clyde Tombaugh (Figure 10.14) was born in 1906 to an Illinois farm family. As a boy he developed an interest in stargazing that was encouraged by both his father and his uncle. The first telescope he ever owned was bought from Sears. By 1925, Clyde was dissatisfied with his store-bought telescope and decided to build one for himself. Clyde's father took a second job to pay for the materials needed to build it. The telescope Clyde built in 1925 was only the first of more than thirty telescopes he was to build over his lifetime. In 1928 Clyde completed the construction of a very accurate

Fig. 10.14 Clyde Tombaugh, discoverer of Pluto, an amateur astronomer, who as a youth built the first of more than 30 telescopes, eventually called to the Lowell Observatory, where he discovered Pluto.

23-centimeter reflector. The mount for this telescope was built from part of the crankshaft from a 1910 Buick and discarded parts from a cream separator! Nevertheless, it was with this telescope that he made the observations responsible for a job offer from the Lowell Observatory. Clyde Tombaugh made very detailed drawings of his telescopic observations of Jupiter and Mars. He sent the drawings to astronomers at the Lowell Observatory, asking for their comments and suggestions. What he received instead was an offer to come to Lowell Observatory to work as a junior astronomer. He accepted the job and joined the search for Percival Lowell's "Planet X", a planet beyond Neptune, which Lowell believed, existed but could not discover. Clyde Tombaugh's job was to photograph one small piece of the night sky at a time. He then had to carefully examine and compare the photos in an effort to detect an unidentified moving point of light that might be a planet. Tombaugh photographed 65 percent of the sky and spent thousands of hours examining photographs of the night sky. After ten months of very hard work, sometimes working through the night in an unheated dome,

Tombaugh discovered the (dwarf) planet Pluto. Clyde Tombaugh died at the age of ninety on January 17, 1997.

Pluto has a very strange orbit. It takes almost 249 years for the planet to go around the Sun and at times it crosses the orbit of Neptune, making Neptune the farthest planet in the solar system. Pluto's orbit is tilted 17 degrees from the mean orbital plane of the other planets — the largest tilt of all the orbital planes. From 1979 until 1999 Pluto was the eighth planet from the Sun and Neptune the ninth. Their ordering changed after that, and does so periodically.

Pluto's atmosphere is extremely thin and the warmest temperature there is about −223°C. Pluto's day is 6.3 Earth days. It rotates in the opposite direction of Earth. Pluto has at least three moons. Charon, the largest, is about half the size of the planet.

10.11. The Voyager Missions

As of August 2005, the twin Voyager 1 and 2 unmanned spacecrafts continue exploring where nothing from Earth has flown before. In the 28th year after their 1977 launches, they each are much farther away from Earth and the Sun than Pluto is and approaching the boundary region — the heliopause — where the Sun's dominance of the environment ends and interstellar space begins. Voyager 1, more than twice as far as Pluto, is farther from Earth than any other human-made object and speeding outward at more than 17 km per second (38,250 miles per hour). Both spacecraft are still sending scientific information about their surroundings through the Deep Space Network. As of July 2003, Voyager 1 was at a distance of 13.3 billion km (which is 88 times the distance between Earth and Sun) from the Sun and Voyager 2 was at a distance of 10.6 billion km. At that time they were so far away that it would have taken a radio signal 12.3 hours for Voyager 1 to send a radio signal to Earth.

The primary mission was the exploration of Jupiter and Saturn. After making a string of discoveries there — such as active volcanoes on Jupiter's moon Io and the intricacies of Saturn's rings — the mission was extended. Voyager 2 went on to explore Uranus and Neptune, and is still the only spacecraft to have visited those outer planets. As part of the current mission, these spacecraft will explore the outermost edge of the Sun's domain — and beyond (Figure 10.15).

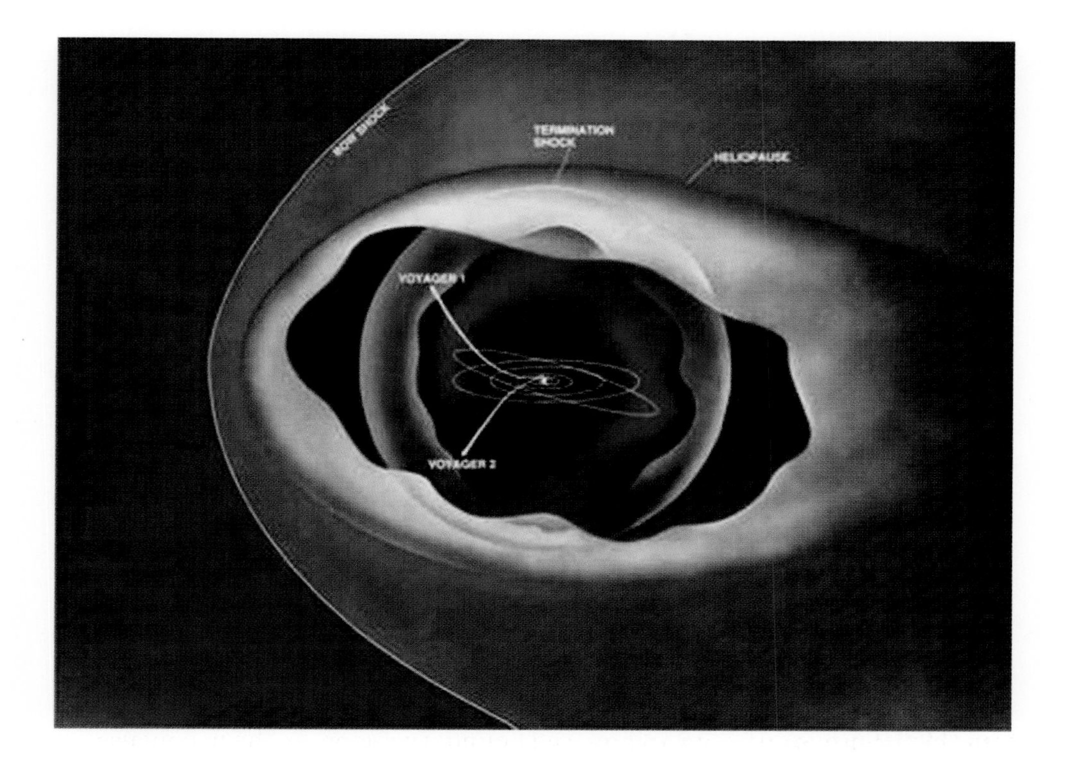

Fig. 10.15 The Voyager missions first explored the far-off planets of the solar system, beginning in 1979 and now extend beyond the Sun's domain into interstellar space. Inner orbits of the planets around our Sun are shown schematically at the center. The present course of the Voyagers into outer space is shown by the arrows. Courtesy: NASA.

10.12. Questions

1. *How can Earth's mass be inferred?*

   By application of Kepler's laws (see page 19) in this case, $M_{\text{Moon}} + M_{\text{Earth}} = R^3/(GP^2)$ (where P is the period or time for one full orbit of Moon around Earth, G is Newton's constant, and R is the radius of the orbit. The mass of the Moon can be ignored in this expression because it is so much smaller than Earth's.)

Chapter 11

New Worlds

This web of time — the strands of which approach one another, bifurcate, intersect or ignore each other through the centuries — embrace every possibility.

Jorge Luis Borges (1899–1986), *The Garden of Forking Paths*

Are we alone in the world? Life seems to be absent from any of the planets of our solar system, at least, from their surfaces. In fact, of the eight major planets, only four are terrestrial, and of these, only Earth has surface water. It wasn't always so. Surface water may have existed on Mars, when it was young, but the Sun's heat and the weak gravity of that small planet caused the evaporation of any water from its surface.

The other four planets are gaseous. But what of other solar systems beyond our own? The night sky sparkles with billions of other suns. From ancient times some philosopher-scientist believed in the existence of other planets around the countless distant suns. Among them was Nicolaus of Cusa.

Nicolaus of Cusa (1401–1464) was the son of a wealthy shipper on the river Mosel, born in Kues, founded by the Romans. In his youth, Nicolaus studied in Padua. There he came into contact with the most progressive traditions of European civilization, which had been revived in Italy with Dante, Petrarch, and Boccaccio. They had started a de facto war against the dogmatic, scholastic teaching that dominated much of the academic life of Europe.

Nicolaus Cusa became a cardinal at the age of 47 and then became the bishop of Brixon. He was interested in geometry and logic and contributed to the study of infinity — the infinitely large and the infinitely small. He looked at the circle, in the mathematical sense, as the limit of a regular polygon and used it in his religious teaching to show how one can approach truth but never reach it completely.

Cusa was a multifaceted individual who thought deeply about many aspects of man's place in the world and the relationship between the individual and the government. In the *Concordantia* he writes;

> *Therefore, since all are by nature free, every governance — whether it consists in a written law, or in living law in the person of a prince ... can only come from the agreement and consent of the subjects. For, if men are by nature equal in power and equally free, the true, properly ordered authority of one common ruler, who is their equal in power, can only be constituted by the election and consent of the others, and law is also established by consent.*

As a philosopher, Cusa is best known for his assertion of the incompleteness of man's knowledge of the universe. He claimed that the search for truth was equal to the task of squaring the circle. In later life he became interested in astronomy and claimed that the Earth moved around the Sun. He also believed that the stars were other suns and that space was infinite. His enquiring mind led him to the notion that an almighty god would not have limited himself to a finite universe, and therefore it must be infinite in extent, and more: he also believed that the stars had other *inhabited* worlds orbiting about them. He may be right!

The man who departed *on the basis of observation* from Aristotelian belief, held for centuries by the Roman Church, was Nicolas Copernicus (1473–1543). His observations were made by sight long before the invention of the telescope in Holland and much improved by Galileo. Copernicus saw that the Earth was not the center of the solar system, but that the planets revolved about the Sun, not the Earth. He had a great influence on Galileo who was eventually forced by the Roman Church to renounce his beliefs and teachings. Galileo fared better than a contemporary, Giordano Bruno, a Dominican Friar, who traveled most of the known world to expound his beliefs gained from his reading of Nicolaus of Cusa and of Copernicus. Bruno was burned at the stake for he would not recant his belief that god would limit himself to only one world populated by life. Galileo's sentence was house imprisonment for the remainder of his life.

11.1. Edmond Halley: The Stars Also Move

Edmond Halley's (1656–1742) name is familiar to us from the comet that he discovered in 1682, and with the aid of his friend's theory of gravity — Isaac Newton — Halley predicted that it would make a return visit to our region of the solar system in 76 years. Halley also noted the distinct possibility that the Earth could be struck by a comet. But most important to us here, in 1710, by comparing current star positions with those listed in Ptolemy's (85–165) catalog, he made the important discovery that the stars are not fixed in position but that they have a slight motion of their own, which is called their proper motion. It is only their great distance from us that gives stars the appearance of being fixed.

A graduate of Oxford, Halley became a member of the Royal Society at the age of 22. From the island of Saint Helena, he catalogued the positions of about 350 Southern Hemisphere stars and observed a transit of Mercury; he urged that the latter phenomenon and future transits of Venus be used to determine the distance of the Sun.

Halley was appointed Savilian professor of geometry at Oxford in 1704, and in 1720 he succeeded John Flamsteed as astronomer royal. At the Greenwich Observatory he used the first transit instrument and devised a method for determining longitude at sea by means of lunar observations. He played an active role in the events and controversies of his time. He both morally and financially supported Isaac Newton. His activities also ranged from studying archaeology to serving as deputy controller of the mint at Chester. He was an integral part of the English scientific community at the height of its creativity.

11.2. The Sun Also Orbits

During the 4.5 billion year life of the Sun, it has circled the Milky Way Galaxy about 18 times. The motion is certainly not perceptible to us. To our eyes we have the impression that the Sun stands still and the Earth orbits about it once a year, and turns on its axis every 24 hours. This much is true for all practical purposes. But the Sun does not quite stand still (quite aside from its long circumnavigation of the entire galaxy). As Jupiter orbits the Sun the mutual gravitational attraction of Sun and Jupiter draw them together, if only slightly. And over the course of Jupiter's year, the Sun, because of the tug of Jupiter, executes a small circle about its mean position. Likewise every other planet perturbs the Sun. And all the planets, if only a little, perturb each other. They do not move about the Sun on perfect circles — that has long been known — but they do not move on perfect ellipses either. Each body effects the motion of every other, according to its mass and proximity. All of this may be very complicated, but Newton's law is at work, and computer programs are able to solve this "many-body" problem. Most important to us here, is to realize that the presence of an *unseen* planet could be detected through its influence on those that can be seen.

11.3. First Planets beyond Our Solar System

Indeed, as Nicolaus of Cusa believed, though purely on philosophical grounds, some suns other than our own do have planets. This has been discovered only recently with sensitive apparatus coupled to computers. Planets that orbit other suns are called *exoplanets* to distinguish them from those of our solar system. Do any of them nourish life? Some intrepid researchers attempt to discern signals from outside our Solar System. That is more problematic than the mere existence of exoplanets, but it is at heart surely the dream of planet searchers. The modern scientific quest for exoplanets, made in our own day, has been rewarded a hundred fold and more.

The search for exoplanets is technically very challenging. After all, planets do not produce light of their own but can only reflect light shining onto them from their host star. They are certainly too dim to be visible, even with the most powerful telescopes, at least with present techniques.

The first exoplanets were discovered in orbit around a type of neutron star known as a *pulsar* (see page 79). Pulsars were first discovered by Jocelyn Bell at Cambridge University in 1967. Embedded in a neutron star is a very strong magnetic field, amplified from the weak field that inhabits stars before they collapse. As the core of the collapsing star shrinks to form a neutron star at the end of the luminous life of an ordinary star, somewhat heavier than our Sun, the flux lines of the weak field are squeezed together, forming a strong field. Generally the poles of this magnetic field will not be aligned with the rotation. As a result, as the star rotates the direction of the magnetic field sweeps out a cone. Along the direction of the field, electromagnetic radiation is beamed. If Earth happens to be intersected by the cone as the pulsar rotates, then Earthbound observers with large radio antennae detect *that* pulsar by "hearing" a beep once every rotation — hence the name pulsar.

It was first noticed in 1991 by Alexandre Wolzsczan of Pennsylvania State University that the pulses coming from a certain pulsar that he had discovered, known as PSR B1257+12, were irregular, but in a somewhat repetitious manner. His data were gathered from a giant radio telescope at Arecibo in Puerto Rico (Figure 11.1). In collaboration with Dale Frail, and after meticulous analysis of the irregularity, they were able to resolve the signal into three totally different oscillating signals. By 1994 they realized that they

Fig. 11.1 The Arecibo radio antenna (telescope) consists of a 1000-foot fixed reflecting surface, made up of 40,000 individual panels, suspended in a natural limestone sinkhole in northwestern Puerto Rico. Incoming rays are reflected back from the surface to two additional reflectors located 450 feet above on the "platform", a 500-ton structure supported by cables from three towers. Credit: NAIC and NSF.

were detecting the beat from a pulsar that was being systematically disturbed by the gravitational push and pull exerted on the pulsar by *three* orbiting planets. The data analysis showed two planets with mass 4.3 and 2.8 times that of Earth's mass and a third that is about two moon masses. Such is the power of careful measurements analyzed by Newton's gravity, even though the stars are 11,000 trillion km away. This was an amazing feat, and has been fully confirmed in succeeding years by other researchers. It is one of the tenets of science: every measurement or observation should be independently reproducible. The first planetary system outside the system had been discovered!

This discovery — the first planets outside our solar system — nonetheless, was not fully satisfying in an anthropological sense, for planets around a neutron star would be uninhabitable by any life form imaginable to us.

Fig. 11.2 Michel Mayor and Didier Queloz of the Observatory of Geneva, where in 1995 they discovered the first planet far beyond the solar system — the first exoplanet. In this photo they are shown at La Silla mountain observatory (Chile) in front of the small 1.2 m telescope with which they have discovered some 40 planets. (Altogether this team has made the first detection of more than 80 planets.) Credit: M. Mayor.

Nevertheless, the discovery raised very interesting physical questions: How did these planets escape destruction by the supernova explosion of the parent of the neutron star? They did, but how?

But what about the existence of planets around suns like ours? Are there far-off *terrestrial* planets that could possibly harbor life? This quest became a focus of intense — and ultimately rewarding search by several groups of researchers.

11.4. The Search for Exoplanets

The first planet outside our solar system that orbits a luminous star like our own was discovered by Michel Mayor and Didier Queloz (Figure 11.2) at the Observatory of Geneva in 1995. It is thought to have half the mass of Jupiter but unlike Jupiter, which lies near the far edge of our solar system, it orbits so close to the parent star that its year lasts only about 4 days. Geoff Marcy and Paul Butler at San Francisco State University confirmed the discovery shortly afterward. Marcy and Butler, themselves, have discovered 75 exoplanets and have studied some of the properties of the suns about which they orbit. They discovered that planets are found more abundantly around stars that are richer in heavy metals than our Sun. Such studies may lead eventually to an understanding of the circumstances surrounding the formation of planets.

The planet discovered by Mayor and Queloz in 1995 orbits a star called 51 Pegasi. It lies near the center of the constellation Pegasus, the Winged Horse, and is located about 50 light-years from our Sun (500 trillion km). Planets that lie beyond our solar system are called *exoplanets*. Since the first discovery, 180 exoplanets have been found (as of April 2006).

The planet itself cannot be seen. Rather, its presence is betrayed by the very slight to and fro motion that the parent star executes as it is tugged by the much less massive planet. The effect is similar to the Doppler shift in the pitch of a train whistle as the train passes by. (More commonly, what is experienced is a car horn blown continuously that changes pitch as the car approaches and then recedes.) In the case of a planet orbiting a far-off star, the Doppler signal is oscillatory as the planet circles its star, alternately pulling and pushing the star ever so slightly toward and away from the observer, as in Figure 11.3.

Of course the Doppler-shift method of planet detection will not work for low-mass planets — their pull on the host star is too feeble to detect. However there is another method that has enabled the detection of planets with masses as small as five Earth masses. The method is referred to as *microlensing*. A nearby star, by its gravitational attraction of the light from a distant star, acts as a lens, as Einstein foresaw, by focusing that light. The focused light is visibly disturbed if a planet should happen to pass through it. The duration of the planet's passage through the focused beam may be very short, and observers have to be very watchful so that sufficient data is gathered in that short time to determine the mass of the planet accurately. A network of observers at various locations around the world facilitates the data gathering.

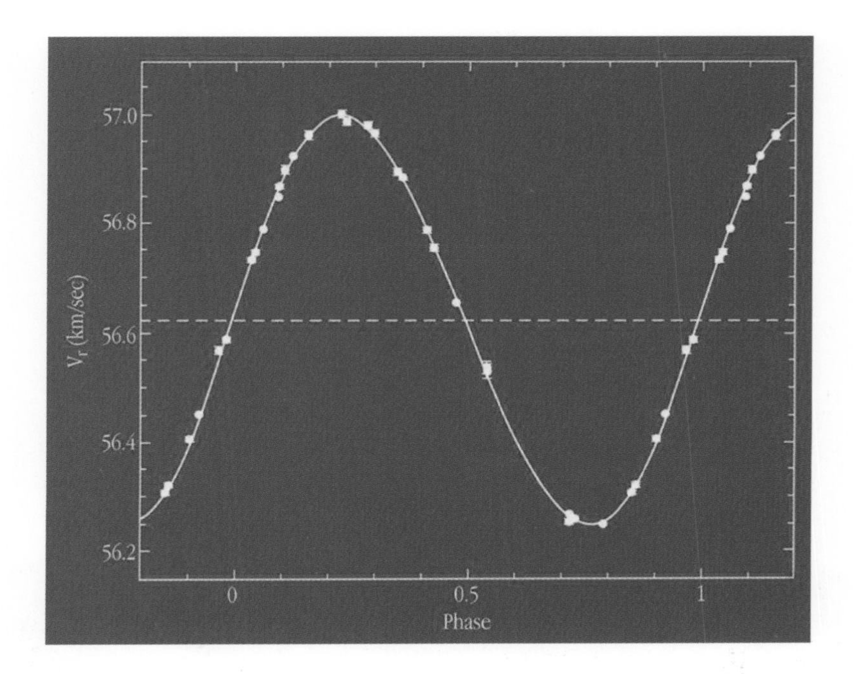

Fig. 11.3 The presence of a planet is seen in the radial velocity curve that shows the oscillating signal received from the parent star (in this case a red dwarf) because of the very slight gravitational push and pull exerted on it as the planet orbits. Gliese 86 (also known as "HD 13445") is seen in the southern constellation Eridanus (The River). It is a bright, rather cool dwarf star, somewhat less massive than the Sun (about 0.79 solar mass). It is also intrinsically fainter than the Sun (about 0.4 solar luminosity). However, since it is quite nearby — about 35 light-years only — its apparent magnitude is comparatively bright and it is just at the limit of what can be seen with the unaided eye. Contrary to most stars with known planetary companions, Gliese 86 contains fewer metals than our Sun, by a factor of two. Credit: European Southern Observatory.

11.5. Search for Life beyond the Solar System

And so the search continues. The imagination of Nicolaus of Cusa in the fifteenth century foresaw that distant suns, far beyond our own would harbor other planets. Based on present exoplanet discoveries, we can safely say that they are as countless as the stars. From our place in the universe, even the closest sun's are far away. This makes the search challenging. Nevertheless, 222 planets circling *nearby* stars have so far been discovered in this new science (as of October 2006). To the present time, none as small as Earth have been seen: the smallest ones so far discovered are about five Earth masses. But this is to be expected because of a selection effect: it is far easier to detect large planets than small ones.

Fig. 11.4 The SETI (Search for Extra-Terrestrial Intelligence) Institute and the Radio Astronomy Laboratory at the University of California, Berkeley, developed a specialized radio telescope array for SETI studies. The new array concept is named the Allen Telescope Array.

As we have described, three methods have been used. One relies on detecting a very small decrease in brightness of a star as a planet crosses between it and us; a shadowing effect. This is the so-called transit method. Another is the Doppler shift method, the means by which the first discovery was made. As a planet circles its parent star, it will ever so gently pull on it, causing the star to alternately move closer to or away from us as it circles its star. The to-and-fro motion is detected by a Doppler shift in the color of the light, redder when the star is moving away, bluer when it is pulled our way. The third is the microlensing method, which uses the gravity of a star to produce a gravitational lens.

What of life beyond our Earth? None has been detected in the solar system. As we have seen, even of the terrestrial planets only Earth is hospitable to life. Is there life beyond our solar system on some planet orbiting a far off star? Nicolaus of Cusa believed so — on religious grounds — that a god who had created our Earth with life on it would surely not limit himself to one inhabited planet. We may never know. Distances are so great that the time for an exchange of signals, should one arrive, may take generations. Nevertheless, a large network of internet connected computers is in operation in the search for extraterrestrial intelligence. Anyone with a computer can participate.

The great Italian physicist Enrico Fermi suggested in the 1950s that if technologically advanced civilizations are common in the universe, then they should be detectable in one way or another. Perhaps apocryphally, Fermi is said to have asked "Where are they?"

From the size and age of the universe we are inclined to believe that many technologically advanced civilizations must exist. Perhaps our current observations are incomplete and we simply have not detected them yet, our search methodologies are flawed and we are not searching for the correct indicators, or other sufficiently advanced intelligence is rare or does not exist.

Bibliography

Berry, M. V., *Principles of Cosmology and Gravitation*, Cambridge University Press, 1976, 179 pp

Finocchiaro, M. A., *The Galileo Affair*, University of California Press, 1989, 381 pp

Freedberg, D., *The Eye of the Lynx*, University of Chicago Press, 2002, 513 pp

Galilei, Galileo (1610), *Sidereus Nuncius* (*or The Sidereal Messenger*), trans. Albert Van Helden, University of Chicago Press, 1989, 135 pp

Galilei, Galileo (1638), *Discourses on Two New Sciences*, trans. S. Drake, 1974, University of Wisconsin Press

Glendenning, N. K. (2004), *After the Beginning: A Cosmic Journey through Space and Time*, World Scientific and Imperial College Press, 208 pp

Glendenning, N. K. (2nd ed., 2000, 1st ed., 1997), *Compact Stars*, Springer-Verlag, 467 pp

Harrison, E. H. (2000), *Cosmology*, Cambridge University Press, 567 pp

Jones, B. W. (2004), *Life in the Solar System and Beyond*, Springer (Praxis), 317 pp

Mayor, M. and Frei, P-Y. (2003), *New Worlds in the Cosmos*, Cambridge University Press, 248 pp

Narliker, J. V. (2002), *An Introducction to Cosmology*, Cambridge University Press, 541 pp

Pais, A. (1982), *Subtle is the Lord* (*Life of Albert Einstein*), Oxford University Press, 552 pp

Rees, M. (2001), *Our Cosmic Habitat*, Princeton University Press, 204 pp

Rubin, A. E. (2002), *Disturbing the Solar System*, Princeton University Press, 317 pp

W. R. Shea and M. Artigas (2003), *Gallileo in Rome*, Oxford University Press, 226 pp

Sharratt, M. (1994), *Galileo Affair; Decisive Innovator*, Cambridge University Press, 247 pp

Silk, J. (2001), *The Big Bang*, W.H. Freeman, 496 pp

Zhi, F. L., Xian, L. S. (1989), (trans. Kiang, T.), *Creation of the Universe*, World Scientific, 180 pp

Zirker J. B. (2002), *Journey from the Center of the Sun*, Princeton University Press, 302 pp

Index